Donald D. Heaton

A Produce Reference Guide to Fruits and Vegetables from Around the World

Nature's Harvest

Pre-publication REVIEWS, COMMENTARIES, EVALUATIONS . . .

"This is truly a gold mine of information on fruits and vegetables, presented simply and in an easy-to-read format. It will be a valuable resource for produce personnel and the food media."

Frieda Caplan
Owner/Founder, Frieda's, Inc., Wholesale Specialty Fruits and Vegetables, Los Alamitos, CA

"Heaton has put together a comprehensive guide to fruits and vegetables from around the world, with extensive histories of their origins and use. He definitely entices the consumer to try new possibilities in the produce department."

Matt Roberts, BS
District Manager of Regional Sales–Specialty Produce, Melissa's World Variety Produce, Brier, WA

"An easy-reading, informative reference guide to an extensive list of fresh fruits and vegetables that are gaining more exposure as we near the global marketplace of the twenty-first century. A must-read for all–from novice to expert."

David H. Bolstad, CEO
Western Mixer's, Inc., Los Angeles, CA

NOTES FOR PROFESSIONAL LIBRARIANS AND LIBRARY USERS

This is an original book title published by Food Products Press, an imprint of The Haworth Press, Inc. Unless otherwise noted in specific chapters with attribution, materials in this book have not been previously published elsewhere in any format or language.

CONSERVATION AND PRESERVATION NOTES

All books published by The Haworth Press, Inc. and its imprints are printed on certified ph neutral, acid free book grade paper. This paper meets the minimum requirements of American National Standard for Information Sciences–Permanence of Paper for Printed Material, ANSI Z39.48-1984.

A Produce Reference Guide to Fruits and Vegetables from Around the World

Nature's Harvest

FOOD PRODUCTS PRESS

New, Recent, and Forthcoming Titles
of Related Interest

Biodiversity and Pest Management in Agroecosystems by Miguel A. Altieri

Winemaking Basics by C. S. Ough

Statistical Methods for Food and Agriculture edited by Filmore E. Bender, Larry W. Douglass, and Amihud Kramer

The Highbush Blueberry and Its Management by Robert E. Gough

Vintage Wine Book: A Practical Guide to the History of Wine, Winemaking, Classification, and Selection by the Sommelier Executive Council

Herbs of Choice: The Therapeutic Use of Phytomedicinals by Varro E. Tyler

Managing the Potato Production System by Bill E. Dean

Glossary of Vital Terms for the Home Gardener by Robert E. Gough

The Honest Herbal: A Sensible Guide to the Use of Herbs and Related Remedies by Varro E. Tyler

Seed Quality: Basic Mechanisms and Agricultural Implications by Amarjit S. Basra

Bramble Production: The Management and Marketing of Raspberries and Blackberries by Perry C. Crandall

Opium Poppy: Botany, Chemistry, and Pharmacology by L. D. Kapoor

Poultry Products Technology, Third Edition by George J. Mountney and Carmen R. Parkhurst

Egg Science and Technology, Fourth Edition by William J. Stadelman and Owen J. Cotterill

Diagnostic Techniques for Improving Crop Production by Benjamin Wolf

A Produce Reference Guide to Fruits and Vegetables from Around the World

Nature's Harvest

Donald D. Heaton

Food Products Press
An Imprint of The Haworth Press, Inc.
New York • London

Published by

Food Products Press, an imprint of The Haworth Press, Inc., 10 Alice Street, Binghamton, NY 13904-1580

Cover design by Marylouise E. Doyle.

Library of Congress Cataloging-in-Publication Data

Heaton, Donald D.

A produce reference guide to fruits and vegetables from around the world : nature's harvest / Donald D. Heaton.

p. cm.

Includes bibliographical references (p.) and index.

ISBN 1-56022-865-2 (alk. paper).

1. Fruit–Handbooks, manuals, etc. 2. Vegetables–Handbooks, manuals, etc. 3. Fruit trade–Handbooks, manuals, etc. 4. Vegetable trade–Handbooks, manuals, etc. I. Title.

SB354.8.H43 1997

641.3'5–DC21

96-49040

CIP

CONTENTS

ABOUT THE AUTHOR

Donald D. Heaton is Produce Specialist at Carr Gottstein Food Company. Located in Anchorage, it is Alaska's largest wholesaler and food store chain. An expert in all phases of produce, he has worked for the company's wholesale division for nearly 20 years, the last four years managing its independent produce accounts and produce business with Alaska's military bases. Mr. Heaton's first involvement with the produce industry came during World War II when, due to a shortage of men, he worked as a produce manager and consultant in the Seattle area, where he also became involved in the creation and management of large fruit stands during the summer months. In 1967, his merchandising ability was honored by an article written in the National Association of Retail Grocer's magazine *NARGUS*. Over the years, Mr. Heaton has attended innumerable produce seminars on produce purchasing, wholesaling, cost control, quality control, merchandising, and displaying of fruits and vegetables.

Foreword

Mr. Heaton has produced an outstanding reference guide for the produce industry to use as an educational tool. The text is interesting, helpful, and well presented. By combining the common fruit and vegetable with the exotic, this book covers a grand spectrum of produce.

The presentation of over 400 produce items offers a wealth of information to the consumer, as well as the tradesman. It fills an informational void created by the rapid expansion of exotic and specialty produce in this country.

Tammy Jerry
Vice President of Perishables
Carr Gottsein Foods Co.
Anchorage, Alaska

Preface

My world of fruits and vegetables opened in the early 1940s, during World War II, when schools let out early so crops could be harvested. The changes since have been enormous–changes not only in the availability and selection of fresh produce, but changes in the way it is grown, packed, and shipped.

Such change also includes the way it is packaged, displayed, and merchandised. From harvesting crops to being a produce specialist, I have had the opportunity and pleasure of working in and observing the produce industry during these dramatic changes.

The produce consumer today has a tremendous advantage over the shopper of yesteryear. The quality, freshness, variety, and presentation of produce today was nonexistent 50 years ago, even in the largest cities. The advent of value-added products, the increasing expansion of genetically enhanced produce, and the fast expansion of exotic and specialty items makes the future of the industry very exciting.

Increased awareness of the value of nutrition and health has altered the way we approach our diet. The old nutritional standard of meat and potatoes has changed; we are now lowering our intake of fat and cholesterol by using greater quantities of fruits and vegetables. Salad bars (unheard of in supermarkets a few years ago) now cater to our increasing demand for a balanced diet. Our anxiety about health has caused us to question the chemicals used in the growth of fruits and vegetables, and this in turn is starting to affect what produce we are buying. Organically grown produce (until recently, it was ignored) is now making its own niche in produce departments.

In the writing of this reference guide, there was some concern about the confusion created by so many names used for the same fruit or vegetable; this concern was also extended to the dilemma made by the use of an individual name to identify several fruits and vegetables.

Adding to this maze was the many trade and hybrip names produced by seed companies, growers, and packers for the marketing of fruits and vegetables. Items such as the mushroom *Agricus bisporus*–previously known as the California brown or Italian brown–gained a new name when in an effort to boost sales, the newer and promising marketing name Crimini was created. It was discovered that when left to grow for a few extra days, the mushroom *Agaricus bisporus* expanded immensely, growing four to six inches in diameter at the cap, consequently creating another marketing opportunity. Thus, the name Portobello was born for this larger mushroom.

Because of the confusion in the names of fruits and vegetables that occur at different points in the production and distribution network, one needs to recognize the inclination to be misled. This is understandable, because as David Weinstein stated in his review, "There simply is not a specific universal set of produce terminology for fruits and vegetables." Understanding this remarkable situation will make the names of fruits and vegetables in this reference guide a little less puzzling.

New methods in farming, harvesting, packing, storing, cooling, shipping containers, and transportation are striving to keep up with the computer age. The availability of exotic fruits and vegetables, which was only a dream yesterday, is now becoming a reality. This reference guide is not only for the novice and consumer, but for any individual who has a desire to learn more about what the world is offering. From atemoya to zucchini, I know you will find something of interest.

Donald Heaton

Acknowledgments

Special recognition should go to the legions of produce professionals, writers, and publishers who amassed much of the information contained in the articles and books researched for this produce reference guide.

Among the many publications researched, articles by Produce Marketing Association; *Packer,* the nation's leading weekly business newspaper for the produce industry; the PMA's monthly magazine *Produce Merchandising;* and their annual *Produce Availability & Merchandising Guide* have proven invaluable.

A special recognition goes to Dean Fahselt, Director of Merchandising for Melissa's (Los Angeles, CA); Laura L. Shovlowsky, Marketing Assistant for Frieda's Finest Specialty Produce (Los Angeles, CA); and Mr. Bob Cobbledick of the Ministry of Agriculture and Food in Ontario, Canada, for taking time out from a busy schedule to help me in my endeavor.

Gratitude without measure goes to my lovely wife, Fern, for her encouragement on this book and for her artistic talent, which made possible the illustrations in this guide.

Introduction

In the years of my involvement with produce, there was always a need for a produce guide that would assist in answering many of the questions consumers asked about various fruits and vegetables on display in markets and produce stands.

The availability of exotic fruits and vegetables unheard of a few years back is now creating a need for accurate information about them. Millions of people have little or no knowledge about exotic produce and many professionals are unable to answer the questions posed by many consumers. It is with this thought in mind that this reference guide was created.

In most cases, this reference guide will provide a short history on the fruit or vegetable, the common and uncommon name(s), a description of appearance and taste, usages, and the time of the year it should be available. This book is an effort to give the consumer and produce novice some knowledge of what the world is offering while also providing assistance to the expert.

To those who would prefer more in-depth information for some of these fruits and vegetables, the three books listed below may be of interest. The books were of equal importance in my research. They are finely illustrated and well written by top professionals in their vocation.

Uncommon Fruits & Vegetables, by Elizabeth Schneider, published by Harper & Row in 1986, is a superb volume and encyclopedia cookbook with over 400 easy recipes that use many of these fruits and vegetables.

Oriental Vegetables, by Joy Larkcom, published by John Murray Ltd. in 1991, is one of the most comprehensive books on oriental vegetables ever written. This is a gardening cookbook that explains the way oriental vegetables are cultivated as well as how they are cooked while also providing each vegetable's history and characteristics.

Vegetables, by Roger Phillips and Martyn Rix, published by Random House Publishers in 1994, not only contains informative text but is magnificently illustrated with over 650 color photographs of vegetables.

Fruits

A

Akee: This tropical fruit was thought to have originated in Guinea, West Africa, and brought to the New World by the slaves that were settled in the Caribbean. Another belief is that Captain Bligh (of *Mutiny on the Bounty* fame) introduced akee to the West Indies along with breadfruit from Tahiti, since its botanical title *Blighia sapida* is named for him. It is now widely cultivated in tropical America and sold in North America where the population includes many people of Caribbean origin.

Small white flowers produce three- to four-inch red-skinned, three-sectioned pods that burst when ripe and split from the stem end to the base end into three sections, exposing shiny, black seeds, which are attached to the base with white or yellowish brain-like fleshy tubes, called arils. The arils also explain the origin of the name "vegetable brain" often applied to this fruit. Cooked arils resemble, in both flavor and appearance, scrambled eggs. The fruit is edible when ripe, but poisonous before pods open naturally and when pods are overripe. No fallen, discolored, or unripe fruit should be eaten. The seeds are also inedible and must be discarded. Akee arils are eaten fresh, boiled, baked, or cooked with seafood. Akee is available from Florida and Central America.

DOMESTIC APPLES

Apple: The apple, whose seeds were first brought to America's shore by the European settlers, is thought to have originated in Southwest Asia. Many believe it is the fruit mentioned in the Bible that was so irresistible to Eve. Its popularity and distribution throughout the nation has been attributed somewhat to folk hero Johnny Appleseed, mom's apple pie, and the rugged pioneers who settled this country.

The state of Washington is the number-one producer of apples in the nation, followed by New York, Michigan, California, Virginia, Pennsylvania, North Carolina, and West Virginia. Although more than 7,500 varieties are grown, a total of 17 popular varieties account for about 90 percent of production. The most popular in order of U.S. production are Red Delicious, Golden Delicious, Granny Smith, McIntosh, and Rome. Two newcomers that will challenge this order are the Fuji and the Gala/Royal Gala. With the advent of controlled-atmosphere storage, apples are now available year-round.

Apple Crab: Crab apples are borne on small trees that bear a profusion of flowers in the spring and, in most cases, attractive fruit in the fall. This tiny apple is round and yellow with a red to maroon blush. It is very tart, not for eating out-of-hand, but good for apple jelly, tarts, apple butter, or garnish. Keep out of direct sunlight. Some varieties available are Southern crab apple, Arnold, Carmine, Siberian, Parkman, Tea crab, Red Jade, Sargent, River, and Red Bud. Available August through December.

Apple Crispin: Usually called crispin in the United States and England, this variety comes from Japan, where it is called Mutsu. It is a cross between the Golden Delicious and a Japanese variety called Indo. A large green apple with a yellow blush, it is an all-purpose apple for eating out-of-hand or for cooking. Available August through June.

Apple Criterion: The Criterion apple is so sweet it is often called the candy apple. In Washington State, this apple is popular at stores and fruit stands each fall. The first Criterion was found in 1968 by Francis Crites in his orchard near Parker, Washington. The chance seedling is thought to be a cross between the Red and Golden Delicious. Criterion inherited the distinctive shape of the Red Delicious, but has a bold yellow color, often with a red blush. The extra-firm, fine-textured flesh resists browning after cutting longer than most apples, making it an excellent choice for salads and fruit trays. It is also fine for baking, and a little lemon juice will enhance its flavor. A late-maturing apple, Criterion is harvested in October and stores well. Since they bruise easily, extra care must be taken when handling. Because they store well, they have the potential for year-round availability. Available late fall through early spring from Washington.

Apple Empire: A mid-season variety, this apple is a cross between the McIntosh and Red Delicious varieties. Developed in 1966 in New York State, its name honors New York, the Empire State. A dark red apple with white flesh, the Empire retains its firmness and juices; flavor is mildly tart. It is a medium round apple with excellent storage life and can be eaten out-of-hand or used in desserts. Available October through June.

Apple Fuji: The Fuji originated at the Tohoku Research Station in Morioka, Japan in the 1930s and was introduced to the Japanese public in 1958. A cross between Red Delicious and Ralls Janet, it has an exceptionally long storage life. In fact, many growers say Fuji's flavor actually improves with age. Typically yellow-green with red highlights, Fujis can also develop a beautiful pinkish blush and sometimes are nearly all red. The Fuji is a late-season apple, often harvested in October, but new strains are being developed that will ripen two to three weeks earlier. This sweet, crisp, juicy apple has a unique flavor and is rapidly becoming a favorite of the apple industry. In the author's opinion, the Fuji is one of the finest apples ever developed. With outstanding flavor, it is superb for eating out-of-hand and excellent for using in salads and cooking. When supply catches up with demand, you can expect them to be available year-round.

Apple Gala: This apple originated in New Zealand from a planned cross between Kidd's orange and Golden Delicious varieties. Selected as the best new variety in 1939, it was released for commercial planting in 1960, and the Stark Brother's Nursery brought it to the United States in 1972. A summer apple that has excellent keeping quality, the gala has a color range from orange-yellow under red striping to nearly solid red. The gala, with its sweet and aromatic flesh and slight cider taste, is probably the most heavily planted new variety. It is best eaten out-of-hand or used in a salad. Available year-round. The other new varieties include Gingergold, Jonagold, Braeburn, Elstar, Jerseymac, Paulared, Cortland, and Idared.

Apple Golden Delicious: Commonly called goldens, these apples are thought to be a nature-bred chance seedling with their origin traced to the Mullina farm in West Virginia around 1912. This single tree bore huge, yellow, sweet, juicy fruit. The apple was introduced to the public by Stark Brother's Nursery in 1916. First used to pollinate Red Delicious trees, goldens gradually came to be appreciated for their unique and outstanding characteristics. Plantings increased steadily through the 1920s, and today they are the second most popular apple in the United States and a favorite worldwide. The skin is light green to yellow, and the flesh has a semifirm texture that is juicy and sweet. An all purpose apple for baking, salads, and fresh eating. Available year-round.

Apple Granny Smith: First introduced to North America in 1958 and promoted heavily by New Zealand growers, the granny has made tremendous advances in consumer acceptance in the past few years. Bright to light green, often with a pink blush for added color, this apple has a tart, crisp flavor. It is juicy and excellent for cooking, salads, and eating out-of-hand. Available year-round.

Apple Lady: Lady is an appropriate name for this petite member of the apple family. It is considered to be one of the "old" apple varieties, and its origin can be traced to France as early as 1628. It was introduced in New York early in the history of this country, and was later grown by such pre-Revolutionary home-owners as George Washington and Thomas Jefferson. This mini-apple, ranging from two and one-half to three inches in diameter, is light green in color with a red blush and a mild-sweet flavor. Excellent for eating out-of-hand. Available from November through December.

Apple McIntosh: An early ripening apple with a two-toned skin that is dominantly red and green, its parentage is unknown. The apple was developed from a chance seedling found in Ontario, Canada, and is still one of the more popular varieties in the United States and Canada. The McIntosh has a light, spicy, and aromatic flavor and is full of juice, and is a very popular variety for eating and cooking. Available September through June.

Apple Red Delicious: First discovered in the 1870s by a farmer in an orchard in Iowa, it has now become the world's most popular apple. The story begins near Peru, Iowa, when farmer Jesse Hiatt planted an apple tree; the tree failed to thrive but a seedling root did. Eight years later, this seedling bore a distinctive apple unlike any other. In 1893, the Stark Brothers Nursery held a fruit fair at Louisiana, Missouri, inviting exhibits of fruits from around the world and offering prizes for the best known and unknown varieties. Hiatt boxed up his apples, "Hawkeyes" as he called them and sent them off to the competition. The apples won a first prize and gained a special name when Clarence Stark tasted the apple and proclaimed it, "Delicious!" Because the tag of origin was lost, the Stark Brother's had to wait another year for another fruit fair to discover Jesse Hiatt and buy the tree of the apple that revolutionized the apple industry. The Red Delicious is identified by five characteristic bumps on its blossom end and a bright to dark red skin with sometimes light red striped or yellowish areas. It is mildly sweet and juicy and is excellent for eating fresh rather than for cooking. Available year-round.

Apple Rome: An almost round apple with a bright, solid-red skin and an aromatic, rich flavor and flesh, the Rome is primarily a baking apple since it holds its shape, texture, and firmness in pies and cobblers. Flavor can be enhanced with a touch of sugar or honey. It can also be eaten out-of-hand. Other cooking varieties include the Jonathan, Stayman, Winesap, Northern Spy, and the Newton Pippin. Most are available fresh from October through June.

EXOTIC APPLES

Apple Bengal: The tree is native to India, and the fruit is often called the bael apple. This fruit is roundish or pear-shaped; two to three inches in diameter; and covered with a smooth, hard gray or yellow rind that contains orange, sweet, aromatic pulp in 8 to 16 cells with seeds. The fruit pulp is used for drinks and conserves, and the flowers are made into perfume in India. It is rarely seen in North American markets.

Apple Custard: It is also called bullock's heart. Native to Central America, the fruit is cone-like or heart-shaped, weighing about a pound, with yellow skin that usually has a red blush when fully ripe. The flesh is creamy in texture, contains many seeds, and tastes similar to the cherimoya, but of poorer quality. It is often used in fruit salads and desserts. Available from Florida and Central America.

Apple Elephant: This tree is native to India, Java, and the Philippines. The fruit is round, acidic in flavor, and about five inches in diameter. It is edible when fresh and is used in curries and jellies. Not available commercially.

Apple Kei: The Kei apple is native to South Africa. The orange-colored fruit is small, roundish, and one to two inches in diameter, and the flesh is yellow with a sweet flavor similar to apricots. The Kei apple is related to the Tropical apricot grown in Florida.

Apple Malay: It is often called the mountain apple and is native to Malaya. The Malay apple is a crispy, red, pear-shaped fruit that is somewhat bland in flavor and whose pulp is similar to the common apple for which it is named. Available from Florida and Hawaii.

Apple Mammee: It is native to tropical America and the West Indies. The large fruit, four to ten inches in diameter, has thick, bark-like brown, bitter skin with pulp that is orange-colored, apricot-flavored, and surrounds one to four seeds. The pleasantly sweet flesh can be eaten raw or cooked. Available as an import.

Apple Otaheite: This tree is native to the Society Islands (Tahiti). The fruit is large, ovoid, and fleshy with tough orange skin; the fruit is two to three inches in diameter. Although unpleasant smelling, the apple-flavored yellow pulp, which surrounds the fibrous core with a spiny seed, is delicious. It is eaten out-of-hand or is used in making preserves. Available October through January.

Apple Star: This fruit, native to Central America, is also known as caimito. The smooth, roundish fruit, two to four inches in diameter, with green or black-purple skin, has sweet white flesh that is star-shaped when cut in transverse sections. Available as an import.

Apple Sugar: Also called sweetsop, it is a popular fruit in India and South America. It resembles the soursop in appearance and the cherimoya in composition. The fruit is round, heart-shaped, oval, or conical, and usually two to three inches in diameter. Covered with a thick, yellowish-green scaly rind and tinged with a powdery-white glow, the interior is creamy white, tender, custard-like, sweet, and slightly acidic in flavor, and is eaten with a spoon as a fresh fruit. The strained pulp mixed with milk makes an excellent drink, and is also used in making desserts. The fruit, a favorite in the West Indies market, contains several long, shiny black seeds. Available August through December from Florida.

Apricot: "Moons of the Faithful" in China and the famous "Golden Apples" of Greek mythology were actually apricots. Native to China, where apricots still grow wild, apricots were next cultivated in Persia. Cultivation then spread to the Mediterranean and eventually to the New World by Spanish settlers. Apricots were first planted in California by the Spanish in the 1700s, and California now supplies about 97 percent of the U.S. market.

Apricots are very susceptible to frost damage during their bloom in early spring. Their production is centered in California, Washington, and Utah. Ripe apricots are one of the most difficult fruits to handle; as a result, new varieties were developed for better shipping. In general, these new types are larger, firmer, smoother, brighter in color, and more acidic in their flavors. Of the older varieties such as Royal Blenheim, Tilton, Derby Royal, and Moorpark, Royal Blenheims set the standard for the classic sweet-tart apricot. Since it ripens unevenly, it is rarely sold in markets. Among the new varieties that include Castlebrite, Flaming Gold, and Katy, came out ahead.

No other fruit surpasses apricots as a source for vitamin A. Apricots are also rich in carotene, potassium, phosphorus, calcium, and vitamin C. Very modest in calories, containing only 231 per pound of product, apricots are ideal as a diet supplement. Until just recently, about 60 percent of the crop was canned; dried apricots represented another 25 percent, with the fresh market using the rest. Available from mid-May through August.

Apricot Tropical: A natural hybrid fruit of Florida, tropical apricots are one to one and one-half inch roundish fruit with velvety-brownish skin and soft, melting flesh with apricot flavor. They are eaten fresh out-of-hand or used in making jams and jellies. Not grown commercially.

Atemoya: A cross between the cherimoya and the sugar apple or sweetsop. It is a member of one of the pudding-like fruits native to tropical America and the West Indies. The atemoya has appeared in the market fairly recently.

The bumpy, tough-skinned, heart-shaped, gray-green tropical fruit has a creamy white, juicy, pudding-sweet pulp with dark seeds. The dense pulp tastes like vanilla custard with overtones of mango, and is excellent for use in fruit salads, sherbets, mousses, and fruit drinks. Ripen at room temperature. Atemoyas are rich in potassium. Available August to November.

Avocado: It is sometimes called the avocado pear or alligator pear. Avocados are native to tropical America, and three ecological races–Mexican, Guatemalan, and West Indian–are recognized. They belong to the family Lauraceae and have been cultivated in tropical America since pre-Columbian times. We borrowed the word avocado from the Spanish. The native Mexican name for this fruit was *ahuactl*, which really meant "testicle."

The avocado has an ancient history. First recorded in Mayan and Aztec picture writings from 290 BC, the avocado has a first-recorded import into Florida in 1833 and into California in 1856. One of the most important fruits in the American tropics, the avocado is grown commercially in many areas of the world including Mexico, Brazil, Australia, Israel, Chile, tropical Africa, California, and Florida.

The first taste of avocados affects people differently; many think it is just a "nothing" flavor, but are usually intrigued enough to come back for a second or third taste. The smooth, rich, nonsweet, buttery texture and nut flavor is one that grows on you. Some people are avocado fans from the first bite.

There are more than 56 varieties of avocados, and these varieties are classified as either summer, fall, or winter. The summer fruit has bright green, smooth, thin skin. The fall and winter varieties are also bright green, but have thicker, rough-textured skin. Of all the commercial varieties, Guatemalans (Haas) garners about 80 percent of the market. Other varieties include Fuerte, Bacon, Zutano, Reed, MacArthur, Pinkerton, Booth 7 and 8, Lula, Waldin, and Gwen.

Half of an average-size avocado contains about 180 calories and is a fair source of vitamins A, C, E, and many minerals. Avocados can be used in salads, dips, garnishes, soups, sandwiches, or just eaten plain with a fork or spoon. Available year-round.

Avocado Cocktail: A seedless fruit of the Fuerte avocado, it is often treated as a vegetable. About two inches long and one inch in diameter, it has a buttery texture and a mild nutty flavor. High in vitamins A, B, C, and E, cocktail avocados also contain potassium, iron, and magnesium. Uses include eating out-of-hand or in soups, salads, sandwiches, and dips. Available from California from December to June.

B

Babaco: Also known as babao or kiwistar, it is native to Ecuador where it is rated as "King" of the exotic subtropical fruits. The babaco resembles a large star fruit, is zeppelin-shaped, and has golden-yellow skin when ripe. The fruit has a taste similar to strawberries, with a hint of papaya and pineapple. When ripe, its skin is a soft gold. Babaco is excellent for eating fresh, for fruit salads, for garnish, and for cooking. Available from New Zealand and California, November through January.

Badea: Native to Ecuador, the badea grows on a vine and resembles a small yellow watermelon with skin so shiny that it looks like it has been waxed. It contains seeds (about the size of watermelon seeds) that are edible, and it makes a delicious drink that tastes something like pineapple juice but without the acid tang. The seeds are served as part of the drink and when chewed, give it an entirely different taste, something similar to grapes. It is not grown commercially and is rarely seen in Northern markets.

BANANAS

Banana: Recorded as early as 327 BC by the armies of Alexander the Great in India, the bananas has been called "the fruit of the wise men." Not only is it the most favored of all fruits, but its nutritional value is exceptional.

The banana was discovered by early Arab traders to India; roots were transported to the east coast of Africa where the plant flourished. In 1482, Portuguese explorers discovered the banana plant there and transported the roots and the African name "banana" back to the Portuguese colonies in the Canary Islands. In 1516, Spanish missionaries brought the banana plant to the islands and mainland of the Caribbean. From there, after traveling nearly halfway around the world, this remarkable fruit was established in Central and South America.

Although more than 150 species are known, virtually all commercial trade is in two varieties: the Gros Michel, a long banana with a tapered tip, resistant to bruising but sensitive to tropical disease, and the more popular Cavendish banana, mild flavored, curved, disease-resistant, but sensitive to bruising.

Unlike other fruits, there is no such thing as a "tree-ripened" quality or flavor. Bananas only ripen satisfactorily after harvesting. A banana grower's family eats bananas that have been removed from the trees while immature, just like the bananas sold in the stores near you.

Bananas are very sensitive to cold and should be stored at temperatures above 56 degrees Fahrenheit. They are rich in vitamins A and C as well as carbohydrates, phosphorus, and potassium. Also, they are low in sodium and contain about 100 calories per large banana. They are excellent for eating out-of-hand and are delicious as snacks, in cereals, fruit cocktails, pies, cakes, and puddings. Bananas are available year-round.

Botanists say the banana plant dates back 1 million years. Bananas are not really grown on trees. Banana plants are giant herbs with enormous leaves. The stem is made up of leaves growing very close together, one inside the other. The fruits form in large bunches weighing up to 60 pounds. Each bunch is made up of 9 to 16 clusters of fruits called hands, and each hand has 10 to 20 separate bananas called fingers. The plant bears just one bunch (stalk) of fruit. When the plant is finished bearing, it is chopped down and another one is planted.

Banana Burro: Also known as a chunky banana, the burro banana has a tangy lemon-banana flavor. The fruit looks like a Cavendish but is flatter and more square. When ripe, the skin is yellow with black spots; the flesh is creamy white or yellow and soft with some firmness toward the center when ripe. They are excellent for eating out-of-hand as well as in fruit salads and other desserts. Available from Mexico year-round.

Banana Red: It is also called the Colorado, Spanish, or Red Cuban. The red banana is a much more squat and heartier banana than the Cavendish. The skin is purple or maroon-red when the banana is ripe, and the flesh is cream-white with overtones of light pink or pale orange. It is sweeter than the Cavendish, less astringent, and the texture is soft when ripe. The fruit is of rather poor quality when eaten raw, but is good for cooking. A good source of potassium, carbohydrates and vitamins A and C, red bananas are available from Ecuador and Mexico year-round.

Banana Manzano: It is a finger-length, yellow banana that is ripe when the skin turns completely black. It has a semidry texture and a flavor that has overtones of apple and strawberry. It is eaten out-of-hand or added to fruit salads and desserts. Available from Mexico year-round. Other banana varieties include the Saba, Cardaba, Blue Java or ice cream banana, Apple or finger banana, and the Brazilian.

Banana Plantain: Used more as a vegetable than a fruit, this member of the banana family abounds in most tropical areas and is the staple food for many of the people from these areas. Plantains are long, curved bananas that are pointed at one end. They are larger, firmer, starchier, and not as sweet as regular bananas.

This fruit (called *platano* in Spanish) is slightly acidic and squash-like in flavor. The skin is thick, and the pulp is salmon-colored. When green, the cooked plantain tastes similar to potato; as it ripens it becomes sweeter. Even when the skin is black, the plantain is ideal for dessert recipes. Preparation for eating includes frying, baking, and boiling. A three and one-half ounce serving contains 119 calories and is rich in vitamin A. Available year-round from Ecuador and Mexico.

BERRIES

Berries: From the first blackberry, red and black raspberry, youngberry, dewberry, boysenberry, loganberry, and Olallie, plump, fresh berries are one of nature's most fragile and glorious fruits. Depending on their sweetness, berries are used differently. Naturally sweet strawberries, blackberries, raspberries, and blueberries are good raw, even without added sweeteners. Gooseberries can be eaten raw or cooked, with or without sweetener. Others such as the tart currants and cranberries are almost always served cooked and sweetened with sugar.

Berry Blackberry: Wild blackberries did not flourish in this country until the forests were cleared, giving the plants enough sun and space to grow well. Blackberry cultivation began around 1825.

There are two classes of blackberries: a trailing, ground runner, sometimes called the Dewberry; and the Bramble Berries that grow on erect plants. Aside from loganberries, which are red, most blackberries range in color from dark maroon to glossy black. Though they look alike, they may differ markedly in flavor and texture. In the northern Scandinavian countries, cloudberries, a close relative to the blackberry, are grown. These are delicious, and a drink is made from them by mixing cloudberry juice with a little brew made from white pine needles.

The most popular commercial varieties of blackberries are Cherokee, Evergreen, Marion, and the Ollalie (a hybrid). Though good eaten out-of-hand, they also make excellent pies, jams and jellies, toppings, and blend well with other desserts. Blackberries are high in vitamin C, and one cup has about 85 calories. Available from July through October.

Berry Blueberry: Wild blueberries have always been a part of the American heritage. From the earliest colonial times, they were the basis of such classic dishes as grunt, buckle, and flummery. The United States and Canada are the largest producers and consumers of this native fruit, with Washington, Oregon, Michigan, Maine, and British Columbia being the largest producers.

Blueberry varieties differ slightly in size and flavor; most are seedless, with a crunchy texture and a sweet, mild flavor that is enhanced by a little lemon juice. Fresh berries should be plump and firm, with a light grayish bloom. Today, Bluecrop is the major commercial variety, followed by Jersey. Other varieties are Rubel, Blue Ray, Bluetta, and Elliots.

The names huckleberry and blueberry are often used interchangeably although they are not the same species. They differ in that blueberries are cultivated, larger, sweeter, and have much smaller seeds. Blueberries are rich in vitamin C, and one cup has about 90 calories. They are used in all manner of cooking, from hotcakes to toppings and other fruit desserts. Available from mid-April through September.

The American Indian was the first to gather blueberries. Eaten raw or baked, they also dried them for use in winter in their soups and stews. Blueberries have been grown and canned commercially since the American Civil War.

Berry Boysenberry: Also called dewberries, they are grown mainly in California and to some extent in Texas and Oregon. Boysenberries are large (one and one-quarter inch long), plump, deep maroon, juicy, and tart-flavored. They were created from a variety of berries: raspberry, loganberry, and blackberry, and can be used like a blackberry. Boysenberries are great in pies, ice-cream, and yogurt. Available June through July.

Berry Cranberry: This plant is native to peat and bog areas of northern latitudes around the globe. Since the first Thanksgiving, Americans have been harvesting these bright red, tart berries from swampy bogs where they grow. Harvested first by hand, then by wooden rakes, cranberries are mechanically picked today, the only berry harvested by machine.

North America is the sole commercial producer of cranberries. Massachusetts and Wisconsin are the largest producers, followed by New Jersey, Washington, Oregon, and western Canada.

Despite more than 100 varieties of cranberries (including lingonberry), only a few are grown commercially: Early Black (a small black-red berry), Howes (a medium-red skin berry), Searles (a deep red berry), and McFarlin, another deep-red berry.

Cranberry sauce was an invention of American Indians who cooked cranberries with honey or maple sugar to eat with their meat. Like currants and gooseberries, cranberries need to be sweetened with sugar. They are good in preserves and relishes and as an accompaniment to meats, especially poultry. They are often prescribed to aid the recovery of people with urinary infections. They do effectively reduce the odor of urine and are included in the daily diets of many incontinents (people without control of urinary functions) for that reason.

Cranberries can easily be frozen. No special preparation is necessary; just wash, put into bags, and freeze. Raw cranberries contain about 44 calories per cup and are very high in vitamins A and C. Available September through December.

In the preindustrial days, children created garlands of cranberries for Christmas tree ornaments by threading them on long strings. Some people still uphold this tradition.

Berry Elderberry: While not grown commercially, this berry has long been used to make jelly, syrups, pies, and homemade wine. The American elderberry (*Sambucus canadensis*) is a common native shrub found along roadsides and at the edge of woods. The berries, grown in large clusters, are dark purple-black in color with a whitish bloom; they are a favorite in the northwest. Green berries and stems are toxic and must not be eaten. Usually available from late July through September.

Berry Gooseberry: Though considered as English as fish and chips, the gooseberry originated in continental Europe. In size, the spectrum runs from pea-sized fruit to the specially cultivated ones of hen-egg size. Gooseberries are not a common sight in the American markets, yet some 50 species are grown worldwide. Requiring the cool, moist climate of the northern states, they do not survive well south of Washington, DC.

In the past few years, there has been a slight increase in commercial quantities coming from New Zealand and a few growers in

California and Oregon. Light green in color, crunchy-textured, and quite tart in flavor, gooseberries are used to make pies and preserves. A new, larger variety, poorman, is sweeter and can be eaten out-of-hand. Gooseberries have about 120 calories per cup and are a good source of vitamin C. Domestic berries are available from May through August, while imports from New Zealand are available from October through December.

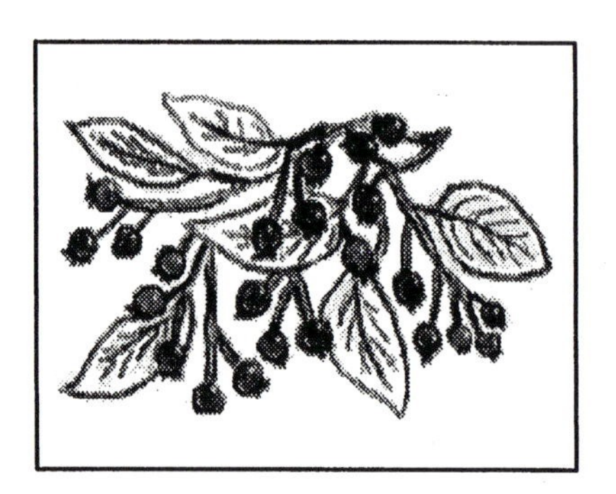

Berry Juneberry: The juneberry is also known as the serviceberry or Saskatoon berry. The little purplish-black berries are the fruit of a small shrub and are highly esteemed in the Great Plains for pies and preserves. The fruit is slightly sweet and juicy with a mild flavor that has been compared to a combination of blueberry and cranberry. Although not grown commercially, they are usually available in late May through July. Other wild berries include the teaberry or wintergreen, bearberry, and the black haw berry, one of the sweetest of all wild fruits.

Berry Keriberry: The keriberry is a newly patented berry, available from New Zealand and Southern California. Neither a blackberry nor raspberry, but called a brambleberry, it is a large berry that is black, sweet, and similar to the blackberry in taste. It is unusual from other berry plants for its ability to produce fruit year-round.

Berry Loganberry: One of the oldest trailing blackberries on the West Coast, it was developed in California and is now grown commercially in Oregon. The berries are large (one inch long), deep red, and quite tart. In some areas, such as western New York State loganberry juice is very popular, but it is usually combined with other sweeter berries for desserts such as pies and jams. Available May through July.

Berry Mulberry: Native to China, they were introduced to North America in the early 1600s, in an attempt to begin a silkworm industry. The mulberry bush of Mother Goose rhymes represents only one form of the plant; most mulberries plants grow to become large trees. Mulberries are shaped like blackberries and range in color from creamy white to lavender, dark red, or black. The berries range in size from three inches in length for some cultivars (Pakistan, for example) to less than one inch for others. The flesh is juicy and has a rich, sweet, subacid flavor. Mulberries are used in puddings, and in making wine, pies and tarts; they are delicious when eaten out-of-hand. Mulberries are not grown commercially.

Berry Raspberry: Raspberries are bramble fruits belonging to the genus *Rubus* and family Rosaceae, the rose family. Though cultivated raspberries were not introduced to the United States until the 1700s, the wild berries were native to this country with the red raspberries growing in the northern states and the black raspberries growing farther south.

There are three groups of cultivated raspberries in North America: (1) red raspberry, a native European berry; (2) black raspberry or blackcap, a North American native fruit; and (3) purple canes with a wide range of hybrids between the three groups, including the golden types. Black raspberries are small and seedy with a distinctive mild-tart flavor. The purple is a hybrid of the black and red varieties, and the golden is a novel variation of the red type but with a milder, more sweet flavor.

Raspberries are the only cane berry that come free of their caps when picked, making them the most fragile of all. About 90 percent of the commercial berries are produced in California, Washington, and Oregon. Michigan and New York also supply some quantities. They are good eaten out-of-hand, for jams and jellies, and for dessert dishes such as Bavarian creme, bombe, and puddings. Raspberries are low in calories and are rich in vitamins A and C. Usually available from June through October.

Berry Strawberry: About 70 percent of America's fresh strawberries are grown in California with the balance coming from Florida, Mexico, and New Zealand. The cultivation of strawberries goes back to the mid-1600s, when early settlers enjoyed the fruits grown by local Indians. But today's strawberries have evolved from five varieties developed at the University of California in 1945. Since then, hundreds of new variations have been developed. Three of the commercial varieties today are Chandler, Pajaro, and Douglas.

California strawberries are bred to be large but the *fraises des bois*, a tiny strawberry of European origin, is now grown on a small scale in California. Strawberries are low in calories, about 55 per cup, and high in vitamin C. They are excellent for eating out-of-hand, and in desserts, fruit salads, toppings, pies, cakes, jams, and jellies. Available year-round.

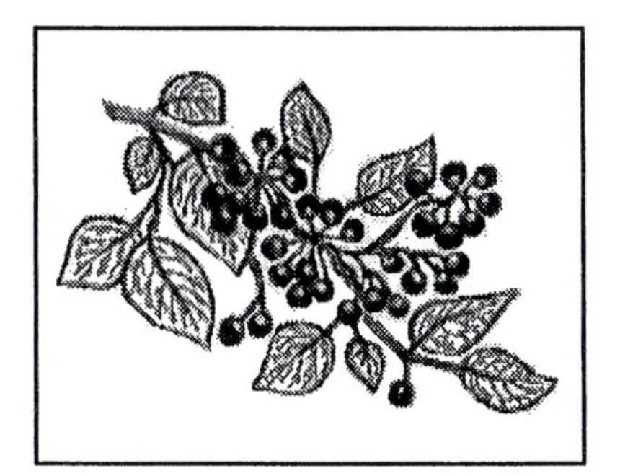

Berry Sunberry: Also known as garden huckleberry, this plant is an improved form of the common weed: black nightshade. The small, black, round berries, grown in clusters and similar in appearance to a currant, are cultivated mainly in North America. The berries are used in the making of jams, jellies, preserves, and pies. Sunberry is not grown commercially.

Breadfruit: It is also known as panapen, pana de papita, and fruit á pain. Native to the South Pacific, few people are aware of its ill-fated first voyage to the New World from Tahiti. When the British decided that breadfruit was the best and cheapest way to nourish the slaves of the British West Indies, they sent Captain William Bligh of the HMS *Bounty* to Tahiti. In 1789, 1,000 seedlings were packed into the hold of the HMS *Bounty* bound for the British West Indies from Tahiti; the rest is history.

Following a dispute over water allocations for the trees versus the crew, a mutiny occurred. Put to sea by his mutinous crew, Captain Bligh survived the best-known mutiny as well as the 4,000 mile journey back to England and then sailed back to Tahiti. Setting sail once again from Tahiti to Jamaica with his ship loaded with breadfruit trees, he arrived there safely in 1793. The slaves refused to eat the breadfruit, and it was not until after the slaves were liberated in the Caribbean (some 60 years later) that it became part of their diet.

Although the name implies fruit, it is used almost exclusively as a vegetable. Breadfruit is solid green, evenly colored, and hard, with large well-developed scales. Weighing from two to ten pounds, it is usually sold whole and partly green. When completely green, the fruit is hard and starchy, like a potato; when slightly ripe, the pulp resembles the texture of eggplant and has a slight musky, fruity flavor that is very bland. Most breadfruit will be speckled with a dried latex-like substance. To remove the white, starchy sap, soak raw cut flesh in cold water for 30 minutes. Breadfruit is usually boiled, baked, and served with butter and gravy, like potatoes, or sliced and fried. Raw green breadfruit has 113 calories per 100 grams and is an excellent source of potassium.

C

Currant Red: Currants were named for Corinth, that immoral city of ancient Greece.* Tart and intensely flavored, red currants are excellent for preserves, meat sauces, and jellies. Currants grow in small clusters or bunches. Select slightly underripe berries for jelly since greener fruit contains more natural pectin. Almost all commercial currants come from Oregon; the season is short, running from early July to early August. Low in sodium and high in vitamin C, currants contain about 62 calories per cup.

Cactus Pear: Also called Prickly Pear, they are also known as Indian figs, Barbary figs, and tuna figs. The Spaniards discovered this cactus fruit being cultivated by the Indians of the Western Hemisphere and took some back to Europe with them. Now, they are found in many areas of the world.

The tough, thorny exterior of the cactus pear covers one of the most ornate fruits you will ever see. Depending on the variety, you may be surprised by a deep magenta color or a beautiful pineapple-yellow shade. The texture of the flesh is similar to that of a watermelon. The magenta-colored pear has a subtle and not too sweet

---

*Corinth, the seat of governmental authority and the leading commercial city of Greece, symbolized in the minds of many people licentiousness and wanton luxury, so much so that the expression "to Corinthianize" came into use to mean "to practice immorality."

flavor, while the yellow-colored pear has a sweeter taste. This fruit is usually served fresh, and the puree is used as a topping or ingredient in pudding. It is also an excellent ingredient for ices and beverages. Available September to December and March to May.

Native-born Israelis are called sabras, *a Hebrew word for prickly pears meaning people who, like the fruit, are tough on the outside and soft and pleasant on the inside.*

Canistel: They are also called eggfruit and ties. The canistel is a three- to four-inch long egg-shaped fruit tipped with a tiny sharp beak. It has a thin, glossy skin covering an orange-yellow flesh that has the consistency of cooked egg yolk. The flavor is similar to cooked sweet potato mixed with heavy cream. Canistels contain from one to three oval-shaped, brown, hard-shelled, glossy seeds about one inch long. The fruit is eaten out-of-hand and is good in fruit salads. Available January and August from Florida.

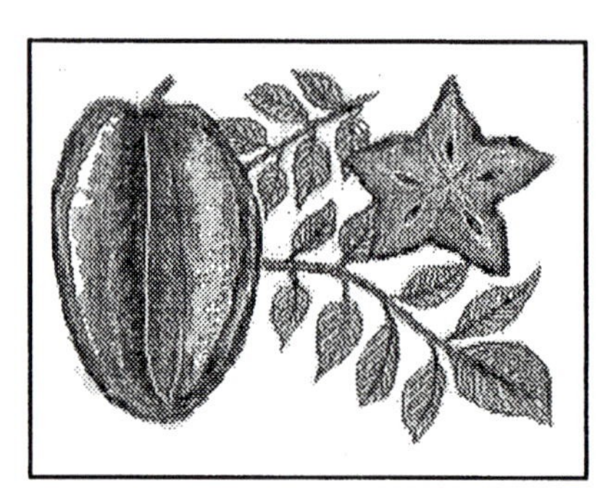

Carambola: Also called the Starfruit, it is no newcomer to cultivation. The carambola is an important crop in Asia, South and Central America, the Caribbean, and Hawaii. Named after a twelfth-century physician, Dr. Averrhoes, carambola is an Indian name for the fruit. It has only been recently marketed here in North America.

The carambola is easily recognizable because nothing else looks like it. A glossy yellow (sometimes white) elongated fruit with five deep longitudinal ridges that give the fruit its star shape when cut into cross sections, it is four to five inches long and about two inches in diameter. Flavor ranges from sweet to tart, depending on variety. Golden Star, grown from seed, has a light, tart taste while Arkin, grown from rootstocks, has a mild sweet flavor. Carambolas are easy to prepare, requiring no peeling or seeding. It is as delicious and appealing in a shrimp sauté as it is in fruit or vegetable salads, or as an ice or mousse. Available August through March from Florida.

Carissa: It is also called the natal plum. The carissa plant is a thorny shrub with plum-like pointed fruit, scarlet red in color, which is excellent for jellies. Available in season from Florida.

Cherimoya: It is also known as a custard apple. Cherimoya trees originated in the South American Andes of Peru and Ecuador and are now grown in this country, usually requiring elevations between 3,000 and 7,000 feet. Another species of the same tree, known as the Custard apple, is of only fair eating quality and is sometimes confused with the cherimoya. Still another species, known as soursop or guanabana, is used for ice cream and cold drinks and is especially popular with Cubans.

These large green, conical, heart-shaped fruits with leathery skins are luscious. The fruit, which are four to six inches long, contain large black seeds and the creamy white flesh, with a custard-like texture, is said to combine the tastes of pineapple, papaya, and bananas. Since the cherimoya is a delicate fruit, it is best served raw or in fruit salads. Available late November through May from South America, Spain, California, and Chile.

CHERRIES

Cherries: By the time agricultural historians began keeping records, cherries were well established in Asia, Europe, and (in wild form) in America. According to Roman writers, this fruit was named after the city of Cerasus in Pontus, at one time part of Greece. The cultivated European varieties were brought to North America by the early colonists, and today the United States is the world's leading producer of cherries.

There are more than 600 varieties of sweet cherries grown in temperate climates. The two most important commercial types are dark red to purple-skinned cherries and yellow-skinned ones. Aside from the sweet cherries, there are more than 300 varieties of sour cherries (pie cherry) and cherry namesakes. Cherries are a fair source of vitamin A, and a cup of raw sweet cherries contain about 65 calories.

The state of Washington is the top producer of sweet cherries, followed by California. Season is mid-May through July. A small supply of sweet cherries from New Zealand is available from mid-November through January.

Cherries are delicious eaten out-of-hand, and are excellent in fruit salads, toppings, short cakes, pies, cobblers, and fruit drinks. A three and one-half ounce serving of cherries contains almost no fat or sodium, but provides about 190 milligrams of potassium and some vitamin A. They also contains about 70 calories per cup (about five calories per cherry.)

Cherry Bing: A leading commercial sweet variety. One of the best fresh-market and commercial canning cherries. It is a large fruit, about one inch in diameter, round and plump with a brilliant red, almost mahogany color. Flesh is purple-red with dark purple juice. Available June through August.

Cherry Black Republican: A medium-size cherry with good quality and distinct flavor. It turns from dark red to black when fully ripe. Available late June through mid-August.

Cherry Lambert: The next big commercial cherry, this variety is a firm, dark red, heart-shaped fruit that matures about a week later than Bings and averages about one size smaller. The fruit is more resistant to late frosts. A Lambert tree is a heavier producing tree. Available late June through August.

Cherry Rainier: This newer variety is delicate both in color and taste. It has cream to golden skin that blushes from red to pink. Flesh is finely textured and clear, and the juice is colorless. Because of their delicacy, Rainiers are packaged in the orchards immediately following harvest. Available late June through mid-August.

Cherry Royal Ann: Also called a cherry Napoleon the Royal Anne has a skin that is light and yellow with a pink blush. The fruit is large, firm, with colorless but flavorful juice. Available June through July.

Cherry Stella: Of good size and moderate firmness, this cherry is similar to the Lambert in shape and color. It has shown resistance to splitting from rain. Other commercial varieties include Chinook, Black Tartarian, Burbank, Chapman, Burlat, Schmidt, and Windsor. Some are available from early May.

Cherry Van: Resembling the Bing in size and quality, this is another popular variety. Its skin ranges from red to dark red, and the fruit has a sweet, juicy flavor. Available late June through mid-August.

Cherry Choke: A small, deep red cherry the choke cherry is very astringent. It is sometimes used for jams and jellies, but is not grown commercially.

Cherry Nanking: Small, bright red, and pleasant-tasting, these cherries are grown on a small tree called the Chinese Dwarf. They are not grown commercially.

Cherry Barbados: Native to the West Indies, northern South America, Central America, Mexico, and southern Texas, this soft, juicy, thin-skinned fruit is crimson when mature and has orange-yellow flesh. It is a good source of vitamin C, and a sweet acidic flavor makes it good for eating fresh, for preserves, purees, and desserts. Available April through November from Florida, Hawaii, and other tropical regions.

Cherry Korean: A small dwarf tree bears cherry-like fruits the size of sour cherries, but much firmer. These cherries are not grown commercially.

Cherry Surinam: Also called pitanga, it is native to Brazil, with the tree seldom exceeding 20 feet in height. The fruit is about one inch in diameter, round, deeply ribbed, with a glossy, bright-red skin. The orange-colored flesh is juicy, tangy, and has one pea-sized round seed, occasionally with two resembling a split pea. Fruit is eaten out-of-hand or in salads, pies, and jellies. Available from Florida.

Coconut: Regardless of where you may live, it is likely that the coconut palm in some way serves you. One of these ways is with the coconut. In tropical lands, there are more than 10 million acres of coconut-palm plantations, with over 600 million trees, not including those grown in villages and elsewhere. Since a tree normally produces about 50 to 100 nuts per year, an annual crop of 30 billion would be a conservative estimate.

The ripened nuts are large and have a thick, smooth, light-colored rind or husk that surrounds the actual coconut and is usually removed prior to export. The nut itself averages about six to ten inches in diameter. If you look at the three holes at the bottom, you will see that it resembles the human face. In Spanish or Portuguese the word *coco* means a grimace or a grimacing face. When selecting coconuts, shake them and listen for lots of liquid, make sure they feel heavy, and check for soft or moldy spots. To easily remove the meat from the shell, first use a sharp instrument (ice pick) to pierce the "eyes" (three small circular spots on one end) and drain the milk from one of the holes. Next, heat the coconut in an oven for 20 to 30 minutes at 325° F. Remove from the oven and wrap the nut in a towel and tap all around the outside with a hammer or mallet to loosen the meat from the shell, then pound it hard to break it open. When the meat is removed from the shell you can peel off the skin with a vegetable peeler. Another method is to freeze the whole, drained coconut and then allow it to defrost. The meat should be easy to remove when you crack it.

Honduras, the Dominican Republic, and Puerto Rico are the primary sources of coconut for North America. Coconuts are very high in fat, up to 60 percent. It is one of the few plant foods that contain mostly saturated fat instead of unsaturated fat. A one and one-half pound coconut yields about three cups of grated meat. Fresh coconut meat per three and one-half ounces, contains 346 calories, 1.7 milligrams of iron, 23 milligrams of sodium, and 256 milligrams of potassium and ascorbic acid. Available year-round.

There are over10 million acres of coconut palms with an annual production of 30 billion coconuts.

Coquito Nut: It is also called a miniature or pygmy coconut. From the Chilean palm, *Jubaea chilensis*, the palm grows very slowly, taking up to 50 years to start producing and remains productive for hundreds of years. The nut is similar in size and shape to a marble, and the skin appears woody, while the flesh is white with a hollow center that holds no milk. It has a nutty taste, a crunchy texture, and should be stored in a high-humidity area to prevent dehydration. To rehydrate and soften product, boil in water for ten minutes. Grind in blender for use like chopped nuts for desserts and baked items. Available from Chile, April to September.

D

DATES

Date: Dates are the fruit of date palms (*Phoenix dactylifera*), trees native to the middle-eastern countries, which come to full bearing after 10 to 15 years and continue to bear for nearly a century, after which date palms gradually decline and, toward the end of their second century, die. The dates grow in immense clusters and are usually harvested in August and September.

Dates are excellent for eating out-of-hand, in fruit salads, bread, cake mixes, pastry filler, and as a substitute sugar. A three-ounce serving contains 230 calories, is an excellent source of potassium, and offers a fair amount of iron and fiber. Domestic dates come from California's Coachella Valley and dry regions in Arizona, such as the Imperial, Yuma, Death, Salt River, and Colorado River valleys. Imported dates, which are grown in the Middle East–mostly in Pakistan–are also available year-round. Although there are several varieties of dates, they can be grouped into three types: soft, semi-dry, and dry. Some of the more popular varieties follow.

The fruit of the date palm was once thought to resemble the human finger; hence, our word "date" comes from dactylus, *the Latin word for finger.*

Date Deglet Noor: Meaning "Date of Night," this date is the major variety grown in the United States. A semi-dry date, it is dark amber in color when fully ripened. It accounts for 95 percent of U.S. production.

Date Halawy: A soft date with very rich, sweet, and distinctive flavor, the Halawy turns golden brown when cured.

Date Khadrawy: A soft date with a greenish cast as it begins to ripen, it has a pleasant flavor, rich without being overbearing.

Date Medjool: The Medjool is a soft, very large, premium date that is amber in color when cured. Because of its large size, the Medjool can be stuffed with a variety of items or prepared as an hors d'oeuvre.

Date Zahidi: The Zahidi is a semi-dry date that is reddish brown when cured.

Durian: (Dorian) Called King of the Fruits, it is native to Borneo and widespread in Malaysia and the southern Philippines. The durian is most unusual, large, ovoid-shaped green fruit with a prickly rind covered with warts and tubercles. Durians grow on a pear-like tree that has oblong, tapering leaves and that is rounded; the fruits tend to grow in bunches.

While the odor of the plant is not so bad, shortly after the fruits are harvested, the smell of this unusual fruit becomes very offensive. When cut open, it reveals five oval compartments, each filled with cream-colored pulp in which there are embedded a number of seeds the size of chestnuts.

The pulp flavor is similar to a rich butter-like custard, highly flavored with almond overtones that bring to mind cream cheese, onion sauce, and brown sherry. So balanced is the flavor that it cannot be described as acid, sweet, or juicy. Because of its smell, most hotels ban it from their menus and some airlines have refused to have it aboard, even when it is tightly packed. Durian is a favorite aphrodisiac in Southeast Asia. It is not grown commercially in the United States but is available in limited quantities from Southeast Asia, Malaysia, and the Philippines.

Another fruit from the same family (Durio) is the *Durian tanah*. This fantastic tropical tree of Malaysia produces numerous large globular fruits that hang on short stalks directly from its trunk and elevated roots. The fruit is a thorny capsule, five to nine inches in diameter, with thick walls that split into five segments, enclosing an aromatic custard-like edible pulp. It is not grown commercially, but may be available from Malaysia or Sabah.

F

Feijoa: It is better known as the pineapple guava, but is unrelated to guava. About two and one-half to three inches long, the small, gray-green, egg-shaped fruit is native to South America. The pale yellow flesh with tiny edible seeds is very sweet, aromatic, and juicy, not unlike the common guava. The flavor is described as a cross between pineapple, quince, and grapes with overtones of mint and lemon. The thick, waxy skin is too tart to eat fresh, but it can be used in pickles or preserves. Feijoas are used to make puree, jelly, fruit butter, chutney, and pies. Available from New Zealand March to June and from California September through January.

Figs: Figs are one of the oldest cultivated fruits. The fig leaf is referred to in the Biblical story of the Garden of Eden. The fig's history in the United States began with fig seedlings brought by the Spanish padres who established California's first Spanish mission in San Diego around 1760. Today, California still produces the country's largest volumes of both black and green figs.

The many varieties of figs usually vary only in color and shape. All have soft, sweet, slightly aromatic flesh that contains numerous edible blossoms and seeds. The leading commercial types of fresh figs are Black Mission, with black skin and delicious pink flesh; Brown Turkey, brownish-purple in color, with richly flavored red flesh; Brunswick, a large dark brown, mild-flavored fruit; Celeste, violet-skinned with rose-colored flesh; Kadota, a large yellowish green fig with violet flesh and excellent flavor; Smyrna, the common imported dried fig; and Calimyrna, a small, amber-skinned fig with a sweet nut-like flavor. Eaten out-of-hand or used in fruit salads, they also make excellent preserves and chutney. Figs are available from California June through October.

G

GRAPES

Grape: One of the oldest fruits known to man, grapes have a long history. While the origin of grape cultivation is obscured by time, scientific evidence indicates vines were propagated as long ago as 4000 BC. For centuries, it was one species, *Vitus vinifera,* an Old World grape. But as civilization increased, seedlings and cuttings spread from Asia Minor to Greece and Sicily. Today, it is cultivated on all continents and islands, wherever the climate is suitable. The original Old English word for this fruit was *winberige* meaning literally "berry of the vine."

Spanish missionaries pushing north from Mexico in the eighteenth century were the first to introduce grapes to California. Soon, the same types of grapes that had been cultivated in other parts of the world were flourishing in the United States.

With more than 8,000 varieties of grapes grown for both wine and consumption, we have to limit our selection to a few varieties that are grown for the fresh market here in the United States.

Raisins are sun-dried grapes and have been used for centuries as a condiment and for commercial purposes. Christian historians say the Israelites paid taxes to King David in raisins.

Grape Almeria: An amber green grape with small seeds, Almerias are firm and juicy, with a sweet, fruity flavor. Available from late September through January.

Grape Black Beauty Seedless: One of the many black grape varieties grown in California, the Black Beauty Seedless is an average-sized berry, with a firm tender skin and a mild, spicy sweet flavor. Available May through July.

Grape Black Corinth: A small red grape that is crunchy and very sweet, the Black Corinth is available early to mid-August.

Grape Calmeria: A medium-size seeded grape, green in color, with an oblong oval shape, thick skin, and a rich, tangy-sweet flavor, it is available October through January.

Grape Cardinal: A large firm, round, frosty-red grape with a mild fruity, slightly tart flavor, it is available late May through August.

Grape Champagne: These grapes are tiny, more like currants in size, and have a very sweet flavor. They are best used as garnishes. Available July through August from California.

Grape Christmas Rose: The Christmas Rose is a new red grape variety that is very crunchy and has a sweet, fruity flavor. Available late September through January.

Grape Emperor: A purple-red grape that is used during the traditional holidays, Emperor grapes are tender, having small seeds and a sweet flavor with a cherry overtone. Available from mid-September through March.

Grape Exotic: A large-seeded grape, very similar to the Ribier Exotic grapes are jet black in color with red overtones, tender skin, and meaty flesh. They are juicy and mildly sweet. Available early June to early August.

Grape Flame Seedless: A round, firm, red grape with a slightly tart flavor and crisp, crunchy texture, it is available mid-June through September.

Grape Italia: A large amber-green grape, crisp and juicy with a rich muscat flavor, it is available mid-August through mid-October.

Grape Muscadine: Also known as scuppernongs, it is one of the few grapes native to North America. Discovered by the early colonists in Georgia and other southern states, it quickly became a favorite fruit of the Southerners. These are large grapes that grow in small bunches and are bronze or purple in color. The skin is very thick and tough, with a delicious pulp that is musky sweet in flavor and similar in taste to the Concord grape (a deep purple grape with a tough skin that is used for wines and jams). It is excellent for eating out-of-hand and for making jams, jellies, and wine. Available September through October from Georgia and Florida.

Grape Niabell: The Niabell is an American-European hybrid seeded grape that looks and tastes like the Concord. Its color is a frosty deep purple. Available August through September.

Grape Perlette: Frosty green, round, seedless grapes with a thick skin and a crisp, mildly sweet taste, Perlettes are usually the first to appear on the market, becoming available in May. Perlettes grow in large, tight clusters.

Grape Queen: A firm, crisp red grape, Queen grapes are juicy and very mild in sweetness. Available mid-August through mid-September.

Grape Red Globe: The Red Globe is a very large, firm grape (the vine boasts berries the size of small plums) with big seeds and a mild sweet flavor. These grapes store well. Available September through January.

Grape Ribier: A large seeded grape, jet black in color with a mildly sweet flavor and a slightly bitter skin, it is an excellent grape for garnish. Available July through January.

Grape Ruby Seedless: The Ruby Seedless is a deep red grape, that is crisp and juicy with a rich, sweet-tart taste and tender skin. Available August through December.

Grape Thompson Seedless: A pale green, oblong grape with a crisp, juicy, refreshingly mild, sweet-tart taste, it is the most favored of all, representing about 40 percent of the fresh grapes market. This grape comes in large sweet clusters and is a good addition to salads and snacks. Available June through November.

Grape Tokay: A red grape, seeded, crunchy, and juicy, Tokays have a fresh, mild, sweet flavor. Available mid-August through November.

GRAPEFRUITS

Grapefruit: The main ancestor of this subtropical evergreen was called the pomelo, brought to Barbados from the Malay Archipelago by a Captain Shaddock. The pomelo, thereafter called shaddock, was quite different from the grapefruit we know today.

The West Indies was probably the point of origin for the grapefruit, evidently a cross between the pomelo and an orange. The grapefruit was one of the latest fruits to be commercialized, with the first shipments starting about 100 years ago. It came to Florida in the 1840s, and about 50 years later when a seedless fruit was discovered, it was propagated to produce the Marsh Seedless variety. Later, hybrids in Florida produced the red and pink varieties.

Florida is the nation's leading producer; along with Texas it provides the majority of the winter crop. Principal commercial varieties grown in Florida are Duncan, Marsh Seedless, Ruby Red Seedless, and Pink Seedless. The leading variety in Texas is the Ruby Red, followed by Ray Ruby, Ruby Sweet, Rio Red Seedless, and Henderson Ruby. Principal varieties in California and Arizona are the Ruby Red and Golden (a hybrid of the Marsh Seedless).

Grapefruit is an extremely popular breakfast fruit; it is useful in fresh fruit salads, as a garnish for fish, and as an ingredient in poultry stuffing. To offer something different, try serving grapefruit with a variety of toppings, such as brown sugar, thinly sliced pepper rings, thin apple slices, and strawberries. One-half of a medium-size grapefruit contains about two-thirds of the RDA of vitamin C, is a good source of potassium and vitamin A, and contains about 38 calories. Available year-round.

The name grapefruit was coined first by explorer John Lunan in Jamaica in 1814 when he noticed how the fruit grew in clusters like grapes.

Oro Blanco: Also called Melo gold, Ora Blanco is a citrus that is a cross between a pummelo and white grapefruit. The Oro Blanco is less acidic than the traditional grapefruit and has a light green to yellow exterior and a golden-colored interior. It can be handled and used the same as grapefruit. Available from California from December through April.

Lavender Gem: It is also called wekiwa or pink tangelo. A cross between the grapefruit and the Samson tangelo, the Lavender Gem looks like a miniature grapefruit in size and shape. It has a delicate, sweet grapefruit flavor. Handle and use the same as grapefruit. Available December through February from California.

Ugli™ Fruit: Also known as uniq fruit, its newly gained popularity began when it was marketed under the name of Ugli fruit. A Jamaican native hybrid fruit that was discovered growing wild in 1914, it is a cross between grapefruit and tangerine and looks like a loose-skinned grapefruit. The fruit is ovoid-shaped and four to seven inches in diameter; however some can grow as large as a small watermelon. The flesh is tart, sweet, and juicy, and the skin color ranges from a deep green to a bright yellow. Despite a rough appearance, Ugli fruit is delicate and should be handled carefully. It is excellent for eating out-of-hand, for juice, and for use in cooking. Since they become bitter with time, they should not be stored for long periods. Available October through December from Florida and Jamaica.

Families in Europe use the bigger Ugli fruit pieces as Christmas decorations or in windows during the winter.

Pummelo: Sometimes called the shaddock or Chinese grapefruit, this fruit originated in the Malaysia/Indo-China area and is cultivated today in China and Japan. Known as citrus grande, the pummelo is the largest of the citrus group and closely resembles a grapefruit. Ranging from the size of a small cantaloupe to almost basketball size, pear-shaped or round, it has a thick, smooth green to yellow skin, which, once removed, leaves a fruit slightly larger than a good-sized grapefruit. The white to pale yellow to pink flesh within is unique, aromatic, sweeter, less acidic, and lacking the bitterness that some grapefruit have. Pummelos are good for jams, jellies, marmalades, syrups and for serving fresh as well. Pummelos contain about 70 calories per cup and are rich in vitamin C and potassium. Available October through February from California.

While most grapefruit have about 12 segments, pummelos commonly have 16 to 18 segments.

Guava: Native to tropical America, guava is now grown in Mexico, the Caribbean, Florida, and California. Depending on the variety, the skin is green or yellow and can be pear-shaped or round like a peach. Guava flesh has a white, red, or salmon color, and the sweet taste ranges from strawberry to lemon to tropical, with a somewhat musky odor. Size is usually two to four inches in diameter.

Guavas are often stewed and used for preserves or the guava paste that is popular in Hispanic cooking. Other uses include jellies, sherbets, fruit drinks, or as a breakfast item. Guavas are low in calories and an excellent source of vitamin C. Available September to January from California and Florida.

J

Jaboticaba: Native to Brazil, this fruit is grape-like in appearance and has a dark maroon skin with a Muscadine grape flavor. Each fruit contains one to four seeds. The fruit grows off the main stem as opposed to most fruits that grow from terminal branches. Jaboticaba is used in fresh fruit salads, sherbets, cobblers, jams, and jellies. Available March through June from Florida.

Jackfruit: A strange relative of the breadfruit tree and native to India and Malaysia, the jackfruit is one of the most unusual and one of the largest fruits known. Weighing up to 90 pounds, about one to three feet long, they are borne all along the trunk of the tree. Although the fruit has an unpleasant odor, the taste outweighs this drawback. The yellowish, soft, flaky, sweet pulp is eaten raw, boiled, or fried, and is delicious in curries. The white seeds are delicious roasted, tasting something like chestnuts. Although not grown commercially, it is sometimes available in the five- to six-pound size in produce specialty departments.

Jakfruit: Native to the rain forests of southern India, the dried fruit is about one-quarter of an inch thick and about the size of a dried peach or apricot. The skin is orange-yellow and has a light, natural sugar coating. Jakfruit flavor resembles iced tea with a touch of spice, and the texture is chewy and tender. Store away from heat, moisture, and light. Fruit can be refrigerated up to one year. It can be eaten out-of-hand, used in salads, or brewed in water for a breakfast drink. Available from Thailand year-round.

Jujube: Also called the Chinese date, it is native to China. This fruit has been grown and eaten for more than 4,000 years. Although botanically unrelated to the date, the jujube resembles it in appearance, texture, and flavor.

Jujube fruits range from marble size to that of a large plum. When ripe, the fruit is the color of mahogany; the skin is smooth, shiny red-brown; and at this stage, the flesh is white and pulpy. Depending on the variety, its apple-prune flavor varies from sweet to acidic. The jujube has a high concentration of sugar (about 22 percent) and one elongated seed. Ripe fruits will keep from one to two months when stored at 50 degrees and up to a month stored at room temperature. Jujubes are eaten fresh, dried, smoked, pickled, candied, and as a butter. They can also be boiled with rice or baked with breads. Usually available during the fall months.

K

Kiwano: Native to Africa, this horned melon is now available year-round from New Zealand and California. Reaching lengths of four to six inches, golden-orange when ripe, the kiwano has white seeds encased in a bright green pulp with a jelly-like texture. This spiky golden cucumber-melon cross has the subtle flavor of cucumber, banana, and lime. Although known for more than 3,000 years, only in the past few years has it been commercially available. Kiwano is usually cut in half and served fresh.

Kiwifruit: Called the Chinese gooseberry or Yang tao, it was later renamed kiwifruit after the flightless kiwi bird of New Zealand. The history of the kiwifruit began in the Chang Kiang Valley of China. It was considered a delicacy by the great Khans who relished the fruit's brilliant flavor and emerald-green color. Introduced to the United States in 1904, it was not until the 1960s that it became established in California as a commercial fruit.

The kiwifruit is unrelated to gooseberries, the former name notwithstanding. The brown, furry skin of the three-inch, oblong fruit conceals an appetizing bright-green flesh near the surface, circles of edible black seeds in the center and a pale green center core. The pulp is soft and pleasantly acid and the flavor is reminiscent of strawberries and watermelon. Cut slices make an attractive garnish on plates, in fruit salads, as cereal toppings, and fruit cups. The fruit contains an enzyme that tenderizes meat; thus it can be rubbed into steaks before broiling. High in vitamin C, a three and one-half ounce portion contains about 35 calories. Available year-round.

Kumquat: Kumquats, a member of the Rue family *Rutaceae*, were cultivated for centuries in China and Japan but were not introduced to Europe until about 150 years ago. The word comes from the Cantonese language, and means "golden orange," first spelled kamkwat. Kumquats look like miniature, elongated oranges, one to two inches long, and grow on a small evergreen tree that seldom exceeds 12 feet in height.

Two varieties are grown commercially, the Nagami (oval) and the Meiwa (rounded). Unlike other citrus, the kumquat rind is much sweeter than its tart pulp. The fruit is entirely edible, the texture is juicy, and the flesh contains tiny white seeds. Kumquats can be refrigerated up to one month. Uses include slicing and adding to fruit salads and desserts, blending and mixing with other fruit in drinks, as preserves and marmalades, and eaten out-of-hand. Kumquats are high in vitamin C and contain about 12 calories per fruit. Available December to May from California and Florida.

L

Langsat: It is also known as lanzone, lanzon, and ayer-ayer. Native to the Malaysian region, it is considered to be one of its best fruits. Like many other Malaysian fruits, it has not been cultivated outside of the Asiatic tropics. The fruit, borne in clusters of five, six, or more, varies in form, but is generally round or ovate, one to two inches in diameter, velvety, straw-colored, and with a thick leathery skin. Each of the five segments of white, translucent, juicy, aromatic flesh contains one to three large seeds. Langsats are usually eaten out-of-hand but can be used in other culinary recipes. Although not suitable for cultivation in California or Florida, some attempts are now being made in Cuba. Available as an import.

Lemon: A native fruit tree of Asia, cultivated for hundreds of years, lemons were transplanted to Europe by the Crusaders, and Columbus is credited with introducing them to the Western Hemisphere on his second voyage. Later, after the introduction of lemons to Florida, wild groves became commonplace of until the freeze of 1894/95, after which there was very little replanting. California then adopted the crop, as lemons were highly prized by the Gold Rush miners of 1849 as a preventative for scurvy (due to high vitamin C content). The word lemon is believed to have come from Asian words meaning "sour or sour fruit."

California and Arizona produce most of the lemons consumed in the United States, as well as about one-third of those used throughout the world. Acidic types are the only ones grown for the fresh market with the Eureka and Lisbon varieties dominating. Florida grows Sicilian types such as Bearss, Avon, Harney, and Villafranco.

Lemons and lemon rinds have many uses including making fruit drinks, garnishing meat and seafood dishes, salt substituting for flavoring water and iced tea, seasoning fish, removing stains, remedying colds, keeping other prepared fruits and vegetables from discoloring, adding zest to cooked vegetable dishes, and as a salad dressing. Available year-round.

Lemon Ponderosa: The Ponderosa is also called the citron lemon by some. The small evergreen subtropical tree, native to India bears fruit that is greenish-yellow when ripe: five to seven inches long; pear-shaped with a rough, warty rind; and one to three pounds. The Ponderosa's juice is of inferior quality, but the rind is delicious as homemade candied peel. Available from Florida.

Lime: Like other citrus, limes came to the New World with the early explorers and seafarers. Both California and Florida produce limes commercially. Limes are divided into acidic and sweet types. Only the acidic limes are grown commercially in the United States. Acidic limes are further divided into Tahiti (Persian), a large fruit type, and the Mexican (smaller) varieties, which are called the West Indian or Key limes.

The fruit, ranging from one and one half inch to three inches in length, dark green to medium green in color, sometimes showing yellow, is a favorite garnish for exotic drinks. Lime juice blends well with other citrus juices, cooler drinks, salad dressings, pies, sherbets, seafood, and melons. A lime with a two-inch diameter contains one-fourth of the RDA of Vitamin C and 20 calories. Available year-round.

Limequat: This fruit, a cross between the kumquat and lime, is green in color and larger than the kumquat. Its uses are as a garnish and in marmalades. The juice mixes well with other fruit and vegetable juices. Available from Florida.

Longan: A Southeast Asian native, it is a close relative of the lychee and very similar in growth habits. The one-inch, lightly pebbled, brownish-skinned fruit with crispy opaque flesh and a single black seed is not as sweet as the lychee, but is pleasantly tart and tastes like a grape. The fruit is eaten fresh, dried, or preserved. Available in late summer or early fall from Florida.

Loquat: It is also known as Chinese loquat, Japanese medlar, Chinese medlar, biwa, or the Japanese plum. This tree is native to China. Loquats are pear-shaped, yellow to orange in color, and two to three inches in length, with a pleasant, sweet flavor similar to apple or pear. The plum-like fruit is considered a delicacy, and its pale yellow-orange to deep orange flesh contains a single round stone.

Available July through August, it is commercially grown in very limited amounts in California, Mexico, Hawaii, and Florida. The most production and consumption of this fruit is in Japan, China, and the Mediterranean region. Because loquats require tree ripening to produce a good flavor, most are canned or preserved with only a few reaching the fresh market. Highly perishable, they require gentle handling and moderate refrigeration.

Varieties include Gold Nugget, Early Red, Advance, Pineapple (which describes the flavor), Oliver, Tanaka, and Premier. Loquats are eaten out-of-hand, served in fruit salads, baked in desserts and pies, preserved, or added to Oriental sauces.

Lychee: Lychee, pronounced LIE-chee, is also known as litchi nuts and lechee. Said to be one of the five finest fruits in the world, this fruit has been popular in Asia for more than 2,000 years where China still grows a substantial amount. Most Chinese lychees sold in the United States are canned, frozen, or dried.

Fruits are grown in large clusters on a medium-size tree; each fruit is about one inch in diameter and one and one-half inches long with a thin bark-like, bright red to red-brown shell. Beneath this shell is a translucent white flesh that contains a smooth brown seed. The flavor is compared to the Royal Ann cherry, and the texture is reminiscent of peeled Muscadine grapes. Lychees can be eaten alone or combined with other fruits in salads, fruit desserts, poultry sauces, and fruit compotes. Lychees are a good source of vitamins B, C, D, and E. A three and one-half ounce serving (about ten fruits) contains 66 calories. Available from mid-June to mid-July.

M

Mango: Cultivated for 6,000 years, mangoes are the most universally popular of all tropical fruits. From Southern Asian origins, both wild and cultivated mangoes have spread throughout tropical regions of the Western Hemisphere. The mango is as common in the tropics as the apple is in the temperate zone, but until recently, this fruit was a rare sight in American grocery stores. Mangoes are now grown commercially in Florida and are imported into the United States. These large, ovoid-shaped fruits with soft, juicy, orange-yellow flesh have a flavor described by some as a mixture of peach and pineapple with a flowery aroma. They range from egg size to some weighing as much as four or five pounds. They may be kidney-shaped, round, pear-shaped, or long and narrow. Mangoes are used in fruit salads, chutney, sauces, purees, or ice cream and are eaten raw or baked.

Today, the U.S. markets are supplied with a large assortment of varieties from the Caribbean, South America, Mexico, Haiti, Puerto Rico, and Florida. The most popular varieties include Tommy Atkins, Kent, and Keitt. Other varieties include the Haden from Mexico, Francescae from Haiti, and Van Dyke from Florida. Whatever variety you choose, the mango is sure to have somewhat stringy, bright orange-yellow flesh clinging to a large, flat seed. Mangoes are high in vitamins A and C. One medium-size mango has about 150 calories. Available May through March.

Would you believe that there is a fruit that designates the passage of time? In Hawaii, kamaainas *(old timers) refer to times as "mango seasons" rather than months or years.*

Mangosteen: Also known as mangostan, it is native to Malaysia, Indonesia, and the Molucca Islands. It is one of the most delicious of the tropical fruits and one of the rarest, due to the difficulty of cultivating and transporting it. It is the fruit of a small tree, and on maturity, the mangosteen reaches the size of a large tangerine and has a brilliant purplish-crimson rind that resembles the pomegranate. It has a thick, leathery rind and contains a pinkish-white pulp divided into segments like an orange with yellow juice and large, flat seeds. The delicious white pulp is soft and melting, with a sweet, slightly acidic flavor.

According to some, the flavor of mangosteen contains the best properties of the mango, pineapple, peach, litchi nut, orange, and grape. Some authorities say it is best when served chilled. It is eaten out-of-hand, made into preserves, or sometimes pureed and used as a topping for ice cream or sherbet. In the East Indies, the pulp may be cooked with rice or syrup. Available July through August from Honduras.

MELONS

Melons: Melons are available in a tantalizing spectrum of colors, shapes, sizes, tastes, and textures. Most melons fall into the broader category of muskmelons. The exception is the watermelon, which is actually a member of the gourd family. Melons originally came from the Orient via Armenia, to Europe, and then to America. Most of the commercial melons grown in the United States are raised in California, Texas, and Arizona; the remainder come from 22 other states. Melons are also imported from Mexico, New Zealand, and Chile. Because of new hybrids, expansion of growing areas, and improvements in care and transportation, most melons are now available year-round. Melons are a good source of vitamin C and those with deep orange or red flesh are rich in vitamin A.

Originating in ancient Persia, melons were among the first fruit grown in Egypt, dating back to 2400 BC.

Melon Cantaloupe: The cantaloupe is a member of the Cucurbit family, which includes cucumbers and squash. The word "muskmelon" properly describes the netted melons, characterized by a network of skin lines or veins that have the appearance of an open-mesh weave, and means the same as "cantaloupe" in America; however, Europeans recognize a clear distinction. Some of the finest cantaloupes grown are from Afghanistan and Armenia where some varieties match watermelons in size. Cantaloupes, including the larger Persian melons, are low in calories and sodium content as well as a good source of vitamins A and C plus three essential amino acids. They are used as appetizers, in fruit salads, snacks, and desserts. Available year-round.

True cantaloupes are not netted and have smooth to rough skin. As Americans use the term (incorrectly), it describes the netted skin and orange flesh varieties including the Persian melons that have green-netted skin.

Melon Casaba: This melon, larger than a cantaloupe or medium-sized honeydew, is a globular winter melon that is chartreuse-yellow at maturity with longitudinal wrinkles that meet at a pointed end. The flesh is thick, soft, juicy, and creamy-white. Two commercial varieties are the Golden Beauty and Sun-gold. Nutritional values are comparable to cantaloupes. For best flavor, choose melons between five and seven pounds. Available May through October.

Melon Chinese Cinnabar: A deep red color, smooth texture, and a sweet, subtle flavor characterize this melon. The Chinese Cinnabar melon is an ideal breakfast fruit and delicious in fruit salads. Available from Florida July through September.

Melon Crenshaw: Like other members of the broader canta-loupe category, crenshaw (or cranshaw) originated in Asia and are a good source of vita-mins and minerals. Slightly smaller than the casaba, it has a golden, netless rind at peak ripe-ness, which is tinged with green while the flesh is salmon col-ored. They are oval in shape with a blunt point at the blossom end; they are easily distinguished from casabas, which are roundish with pronounced ridges, because of their barely detectable ridges. For best flavor, choose melons between five and seven pounds. Avail-able May through October.

Melon Juan Canary (Canari): First planted in the United States in 1972, this melon is canary yellow and oblong; the flesh is white, very sweet, fragrant, and juicy, with a tinge of pink around the seed cavity. Ripeness is determined by softness at the ends. For best flavor, choose melons between four and five pounds. Available May through October.

Melon Honeydew: Historical information is sparse; the name was given by a pioneer grower in Colorado who had been sent seeds thought to have been those of the White Antibes variety grown in southern France and Algeria. Honeydews are a good source of vitamin C, potassium, and trace minerals.

Honeydews are unsurpassed among melons for their fine flavor. A good honeydew is very juicy, with crisp, pale green, honey-sweet flesh. A ripe honeydew has a nice aroma, a creamy skin color, a slightly soft blossom end, and the feel of loose seeds when shaken. For best flavor, choose melons that weigh five or more pounds. Available year-round.

Melon Kavamelon: A member of the muskmelon category and about the size of a large honeydew, the kavamelon has a taste similar to the honeydew and the texture of a watermelon. For best flavor, choose melons in excess of five pounds. Available June to December from California.

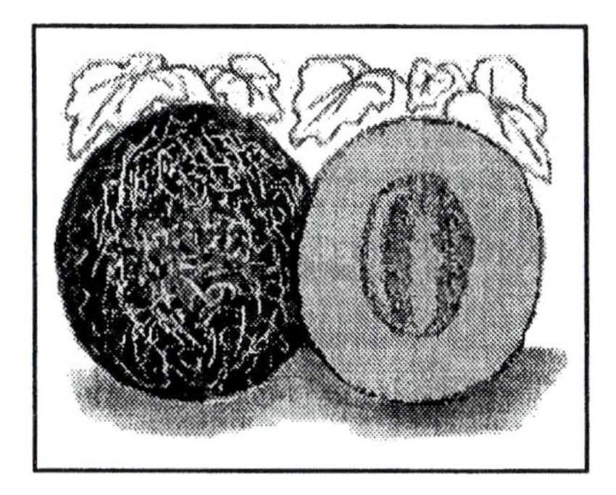

Melon Persian: This type of cantaloupe is about twice the size of the ordinary cantaloupe by weight and has a close brown netting over a green background color. The most round of all melons, Persians have thick salmon-colored flesh that is crisp sweet, and very juicy. For best flavor, choose melons in excess of five pounds. Available May through October.

Melon Orange Flesh Honeydew: In color, taste, texture, and aroma, this small melon resembles a cantaloupe more than a honeydew. When the melon is ripe, the rind turns from white to light salmon pink. Available May through October.

Melon Santa Claus: Sometimes called the Christmas melon, this melon resembles a small watermelon with a mottled yellow and dark-green rind. The pale green flesh is not as sweet as many other melons and tastes similar to honeydew. For best flavor, choose melons in excess of five pounds. Available May through October.

Melon Sharlyn: Sharlyns are oval-shaped melons that are especially sweet and fragrant, and the flesh is creamy-soft with a yellowish tinge. Excellent when used in fruit salads or as a breakfast treat, sharlyns are available June through September.

Melon Pepino: Also called the melon pear, melon shrub, and mellofruit, this subtropical fruit is now imported from New Zealand. To look at the pepino, you would never guess it was a melon–only the flavor gives it away. Native to South America, this two to four inch, teardrop-shaped fruit is also grown in California on a small-scale commercial basis. This exotic fruit is marked with purple and greenish-yellow stripes, the flesh is pale yellow-green to yellow-orange with a melon-like texture, and the flavor is like that of a cantaloupe crossed with a honeydew. Pepinos are low in calories and high in vitamin C and potassium. Available from February through December.

Specialty Melons: French Afternoon, Charentais, Prince, White Breakfast and French Breakfast are newcomers. There are also two newcomers from Israel: the extremely sweet Galia melon and the delicately sweet Ha-Ogen.

Monstera: Also known as ceriman or Mexican breadfruit, the monsterea is a native of tropical American forests. It is a member of the Arum family (familiar as the split-leaf philodendron). Its flower, consisting of a central spadix surrounded by a hood-like spathe (like the antherium), is the start of this very unusual and amazing fruit.

Twelve to 14 months after the flowers appear, the nearly foot-long spadix (the fruit) begins to ripen, and the hexagonal pieces begin to split from each other and from the creamy pulp within. Only when fully ripe is the fruit edible. When ripe or slightly overripe, it is delicious. The grayish-white pulp forms closely packed wedges similar to the design of pineapple scales. Do not refrigerate. As the fruit ripens, the platelets will loosen or pop off; then you can fork out the flesh or pulp for use. The soft, juicy flesh tastes like fresh pineapple and ripe banana. Fresh, it can be added to milkshakes or used as a chutney base or a glaze for pork or chicken. It also makes a delicious spread for bagels or toast. Available August to September from Florida.

N

Naranjilla: Native to Ecuador and the Andes, naranjillas (which means "little orange" in Spanish) look like little oranges about the size of tangerines. Naranjillas have smooth shiny skins like tomatoes, but they are covered with minute fuzz, somewhat like a peach, and the fuzz is hard and brittle. Used in drinks, they have a flavor similar to a mixture of pineapple, orange, and apple with a slight overtone of tomato juice. They are rarely seen in the North American markets.

NECTARINES

Nectarine: The nectarine has been recognized as a distinct variety of fruit and is not as some suppose: a fuzzless peach or a cross between the peach and plum. In fact the nectarine grew wild in Asia more than 2,000 years ago. The nectarine of today bears little resemblance to those grown in the past.

Arriving to the United States from Europe by early immigrants, it was introduced to California about 130 years ago. Nectarines have risen in popularity over the past few years, and some growers predict that they will surpass peaches in the near future.

With a pit similar to the peach, the nectarine has a thin skin, which is reddish-yellow to almost crimson. The sweet, succulent, juicy flesh, which may range from white to yellow, has a full, rich flavor similar to the peach. The nectarine is excellent for eating out-of-hand, in fruit salads, as a topping for cereals, as garnish for poultry, fish, or ham, or in cobblers, pies, and various other desserts. A five-ounce nectarine contains only 65 calories, three grams of dietary fiber, 15 grams of carbohydrate, 271 milligrams of potassium, and is a source of vitamins A and C. Nectarines are virtually free of sodium and fat and contain no cholesterol.

More than 150 varieties are grown in California. California accounts for nearly 100 percent of the nation's production, and the fruit is usually identified in type (whether clingstone or freestone) and color (whether white or yellow flesh). A few of the varieties that dominate the market now follow.

Nectarine Fantasia: Freestone with 50 to 75 percent red blush. One of the more popular varieties. Available July through August.

Nectarine Firebrite: Semifreestone with 50 to 90 percent red blush, usually averaging 75 percent. Available June to July.

Nectarine Flamekist: Clingstone with 40 to 70 percent bright-red blush on sunny-yellow background. Available August through mid-September.

Nectarine Flavortop: Freestone with 50 to 80 percent red bush. Aptly named for its sweet-tart nectarine flavor. Available late June through July.

Nectarine May Glo: Clingstone with red blush over 50 to 85 percent of its surface. Available May through mid-June.

Nectarine May Grand: Semifreestone with a 50 to 90 percent red surface blush. Available May through June.

Nectarine Red Diamond: Freestone with 90 to 100 percent red when ripe. Available June to July.

Nectarine Red Jim: Clingstone with up to 90 percent variegated red blush. Available early August to late August.

Nectarine Royal Giant: Clingstone with 40 to 70 percent red blush over greenish-yellow background. Available July through August.

Nectarine September Red: Clingstone with 60 to 90 percent variegated red blush. Usually the last major variety of the season. Available from third week in August into October.

Nectarine Spring Red: Freestone with 70 to 95 percent red blush. Available June to July.

Nectarine Summer Grand: Freestone with 60 to 95 percent red blush with golden-butter background color. Available late June through July.

NUTS

Nut Almond: No one knows exactly where the almond originated. Undoubtedly, *Prunus ulmifolia*, the wild species of China, contributed to our present-day nut. Cultivated throughout recorded history, it is one of the two nuts mentioned in the Bible (the other being pistachio). Virtually all are grown in California where the Nonpareil is the dominant variety, comprising about 50 percent of the crop. These are followed by Mission, NePlus, Peerless, Merced, and other varieties.

Refrigeration preserves almond quality for very long periods. The protein content is extraordinary (18.6 percent). Almonds are a good source of thiamin and riboflavin, several minerals, and unsaturated fat, 20 percent of which is linoleic, an essential fat the body needs and cannot synthesize.

They are eaten out-of-hand as a holiday season favorite; used to garnish puddings, candies, and ice cream; served with peas, green beans, fresh fruit salads, Waldorf salads, and fish; and used as an ingredient in fruit cakes, rice dishes, chicken dishes and Lobster Newberg. To blanch shelled almonds, pour boiling water over the nutmeats and let stand for approximately five minutes. Then place them in cold water for a few minutes and drain. The skins will slip off easily. Almonds are available year-round.

Nut Brazil: It has now been renamed Amazonia but is also called paranuts, cream nut, butter nut, and (in Europe) American chestnuts. Amazonian nuts are gathered from the dense rain forests of Brazil and are a standard part of the in-shell nut assortments offered in supermarkets.

Grown on trees called *almendros* which tower from 100 to 150 feet high, these woody and wrinkled shells are borne in clusters of 10 to 25 inside a globular fruit called coco (a hard, spherical shell about four to six inches in diameter). Each nut is kidney-shaped with a triangular cross section and arranged like segments in an orange. The kernels are similar in taste to coconut meat. Production remains concentrated in Bolivia and Brazil from whence they originated. Storage and handling requirements parallel those for almonds. Available year-round.

Nut Cashew: Cashews are a member of the Anacardiaceae family, which includes poison ivy, poison oak, poison sumac, the mango, pistachio, and the smoke tree. Native to the West Indies, Central America, Peru, and Brazil, the edible nut is the seed of a tropical evergreen tree. Transplanted by the Portuguese in the East Indies in the sixteenth century, the cashew was later established on the eastern coast of Africa. Today, the leading cashew-producing countries are India, Brazil, Nigeria, and Mozambique.

The kidney-shaped nut, the size of a large bean, is borne on the blossom end of a yellow or orange pear-shaped edible fruit called a cashew apple. The sour fruits can be eaten after processing and are used in making condiments. They are also fermented and used in making wine.

The kidney-shaped ovary (nut) has a hard double shell, and between the shells is an extremely caustic black oil that has to be burned off before the nuts can be touched. The oil is used in plastics and varnish industries. After the outer shell has been burned off, the kernels are then boiled or roasted again, and the second shell is removed, freeing the nut. The nut is then used as a food, a source of food oil, or as a flavoring. The tree yields a milky gum that is the basis of a special varnish used to protect books and woodwork from insect damage. These nuts are only available in processed form. Available year-round.

Nut Chestnut: Chestnuts are traditionally a Thanksgiving and Christmas holiday special. Roasted until the steam inside burst their leathery jackets, chestnuts have a unique nutty flavor and mealy texture.

Until 1904, chestnut trees grew on the East Coast, from Maine to Florida. In that year, a grower on Long Island planted saplings from the Far East, importing at the same time a fungus disease (blight) that destroyed all American chestnuts by 1940. Before the introduction of the potato from the New World to Europe, chestnuts and turnips were the staples in the diet of the poor.

The best varieties bear heavy crops of well-flavored nuts that are much larger than the American chestnuts were. Of the three varieties available–Chinese, Japanese, and European–the large European chestnuts are the ones usually associated with Thanksgiving. Japanese, Italian, and French people consume the most chestnuts; however, they are also consumed in large amounts in England and the United States as well. Chestnuts are popular for poultry stuffing as well as for eating roasted. Available through the U.S. winter holiday season.

Nut Filbert: Also known as hazelnuts, they are a standard part of in-shell nut offerings during the U.S. fall and winter holiday season. This nut was originally named philibert after an obscure Burgundian saint. These marble-size, round, amber nuts yield a single kernel with a sweet flavor and firm texture. Wild trees are found throughout Europe and much of North America. Turkey once was the main source, but now almost all commercially grown filberts are produced on the western slopes of the Cascade Mountains in Oregon and Washington.

Ground filberts make a delicious topping for desserts, cakes, tortes, and souffles. The whole nuts are used in making candy as well as for eating out-of-hand. They make an excellent nut butter. The 13 percent protein content and 5 percent sugar content are low among major nuts, while the fat content is one of the highest.

Nut Macadamia: Native to Australia, macadamias are now being grown commercially in California and Hawaii. The first macadamia tree was planted in Hawaii about 1890 and grew so well that it is now grown commercially on the Hawaiian islands of Oahu, Maui, Kauai, and Hawaii. The evergreen trees, growing as high as 60 feet, bear thick husks that surround hard-shelled round nuts. Since the shells are extremely hard to crack, macadamia nuts are usually marketed shelled, either in tins or jars. The flavor of the kernel is similar to the Brazil nut, only milder and more delicate. They contain 70 percent fat and are a good source of minerals and vitamins. They can be eaten out-of-hand, ground for topping, or used in candy. Available year-round from Hawaii.

Nut Peanut: The peanut is also called American groundnut, or goobers. About 62 percent of this popular U.S. product is grown in the Southeast. Three types are the most commonly grown: Virginia or Runner, Spanish, and Valencia.

Available shelled or in-shell year-round, these nuts have an hourglass-shaped shell and usually contain at least two nuts. Sold in-shell, sometimes raw as well as roasted, peanuts are available in good volume throughout the year, unlike other in-shell nuts that are so strongly associated with the Christmas season.

Peanuts are not nuts and they do not grow on trees. Instead they grow underground on a leguminous plant related to peas. Their protein content (30 percent) is much higher than that for tree nuts. They originated in South America, probably Brazil, although Africa or China are often credited with this contribution to humankind. Peanuts are eaten out-of-hand, boiled, used in candies, salads, and dressings. Additionally, peanut butter is the most favored sandwich spread for children as well as adults.

Nut Pecan: Produced mostly in the southern United States, pecan varieties include Stuart, Wichita, Desirable, Schley, and Western. Before being sold in their light brown, smooth, oval shells, they are washed and waxed. The Indian word *pecan* describes one of America's favorite nuts, which contributes over one-third of supermarket in-shell nut volume. Pecans are not only eaten out-of-hand but make an excellent pie and are used in making candies and other baked foods. Pecan pies rate next to mom's apple pie as America's favorite baked desert. Easy to store, they have a good shelf life at room temperature and up to a year in refrigeration. Available year round.

Nut Pistachio: In the history of pistachios is royalty, perseverance, and pride. Pistachios date back to the Holy Lands of the Middle East, where they grew wild in high desert regions. Legend has it that lovers met beneath the trees to hear the pistachios crack open on moonlit nights for the promise of good fortune. The royal nut was imported by American traders in the 1880s, primarily for the U.S. citizens of Middle-Eastern origin. California entered the world pistachio market in 1976 and in less than a decade grew from the fifth largest producer to the second largest producer in the world.

Pistachios grow in grape-like clusters on the trees. The nut itself develops after the shell has already formed. The enlarged nut then pushes on the surrounding shell to cause a natural split. The hull, or protective covering over the shell, remains intact, but removes easily when pinched after ripening; nuts also fall easily when the cluster is shaken, indicating the crop is ready for harvesting. California pistachios are mechanically harvested, then rushed to the processing plant, where they are hulled, dried, and sorted, later; the nuts are roasted, salted, and sometimes dyed. Pistachios are excellent as a snack, for use in ice cream, puddings, and many gourmet dishes. Its uses are expanding still as its popularity gains with the increased production. Available year round.

Pistachios were a favorite of the Queen of Sheba, who hoarded the entire Assyrian supply for herself and her court.

Nut Walnut: California produces 80 percent of the September to November domestic supply. Major walnut varieties are the Hartley, Franquette, Payne, and Eureka. The English walnut has a golden brown shell that is slightly rough and hard. Kernels are cream-white and clean. Walnuts are available both in the shell and as shelled kernels.

A native of Persia, the English walnut was a great delicacy in ancient Greece and Rome. The Greeks and Romans attributed to it all kinds of magic healing powers. Later, the English adopted the walnut with the same enthusiasm except it was for the beauty of the trees, the value of the wood, and then, the usefulness of the nut meats.

Keeping qualities are excellent at room temperatures. The kernels yield a high proportion of unsaturated fatty acids and are a good source of iron and vitamin B. Walnuts are eaten out-of-hand and used extensively in salads, candies and baking. Available year round.

Nut Black Walnut: This native-American tree has long been prized for its richly flavored nuts. Its roots produce a substance that is injurious to some plants, so the trees should never be planted near shrubs, vegetables, or flowers.

The black walnut has a rough, tough, hard shell and is related only by name to the English walnut. Their much stronger flavor and shelling problems make them a specialty item in most markets. Their flavor is considered superior to the English walnut, especially for baking purposes.

The American Indians used to crush the walnut husks to gather the oils contained therein. The oil was later released into a slow-moving stream to gather fish–the effect of the oils paralyzed the fish, much the same as rotenone.

ORANGES

Oranges: Oranges are believed to be native to southern Asia. The orange traveled to Syria and Persia and then to Spain and Portugal. From there, it traveled with Columbus to the New World and the West Indies. Early Spanish explorers took it to Florida, and Spanish missionaries took it to California. The word "orange" stems from Arabic and Persian terms for the fruit.

Sour varieties have been cultivated since well before the Middle Ages, with the sweet ones appearing only in the fifteenth century. The Valencia variety originated in the Azores; the seedless navel was bred first in Brazil, and the popular pineapple (named because of its aroma) was developed commercially in Florida. The temple orange is believed to have originated in Jamaica.

Calories per medium-size orange are about 70; oranges are a good source of vitamin C, four of the B vitamins, and a host of important minerals. They are used in fruit salads, lunchboxes, sauces, custards, cake frostings, and many other dishes. Orange juice is a popular breakfast drink, and some recipes recommend the use of orange juice in basting turkeys and other poultry.

Orange Valencia: After bananas and apples, oranges are the most popular fresh fruit. Valencia is the main variety in Florida and California with about 50 percent of the crop coming from California. Valencias are roundish, golden orange in color, slight greening around the stem end, and they contain seeds. The majority of the crop is utilized for juice.

Hamlin and Parson Brown are early Florida Valencia varieties, available from October through the early winter; they have glossy, thin skins. These are followed by the midseason Pineapple variety, also with thin skin and numerous seeds.

Orange Navel: Navels are distinguished by a navel formation on the blossom end. They are about 10 percent of the orange volume, coming mostly from California in late November through June. Their thick skin peels easily, making the seedless navels the preferred orange for eating out-of-hand. Varieties include the thin-skinned Bonanza, followed by the early Washington, and then the much-preferred late Washington.

Orange Temple: Temple oranges come from Florida, although some are grown in California and Arizona. They are tender and sweet and require careful handling. Since they peel almost as easily as a tangerine, they are popular as a fruit snack. Oranges benefit from refrigeration as long as the temperature is not below 40 degrees. They are nutritious and used for juicing, garnishing, and everything from breakfast plates to roasted poultry. Available year-round.

Orange Blood: Also known as the pigmented orange or moros, it was brought to America by Italian and Spanish immigrants in the 1930s. The fruit is small to medium size and has a smooth or pitted skin and a deep red flesh. Light, temperature, and variety will affect its color. The juicy red flesh has a rich orange flavor with raspberry/strawberry overtones. Sweetness depends on variety and growing area, and the fruit is less acidic than common oranges. Blood oranges are a good source of vitamin C and are low in calories. Available from California from December through March.

Orange Chinese Navel: A specialty orange grown in California with a sweet flavor and low acid content, it is eaten fresh or used in fruit salads and desserts. Until recently, most were exported to Hong Kong.

Orange Seville: It looks like a sweet orange, but its flavor can be sour and bitter. The Seville orange is used in marmalades, drinks, and food flavorings. The Seville-orange is available late August through September from Florida.

Orange Shamouti: It is also known as the jaffa orange. A popular eating orange in Europe and grown commercially in the Mediterranean area, this fruit, which is sweet, and juicy is easy to peel and has no navel on the blossom end. Its uses include eating out-of-hand, peeled and sliced, and chunked for fruit salads or as a garnish. Grown in California, Texas, and Arizona, it is sometimes imported midwinter.

MANDARINS

Mandarins: Mandarins are available October through May. Peak period is November through January with about 82 percent of the crop produced during this period. Florida, California, Mexico, and Arizona provide the bulk of supplies, with Florida serving as the main supplier October to January, and California leading late winter and early spring.

Mandarin fruit is smaller than oranges and has a loose outer skin that is easy to peel. They are delicate and require constant refrigeration to maintain quality. A good quality mandarin will be heavy for its size and will be deep orange or almost red, depending on the variety. A puffy appearance and feel is normal. They are popular ingredients in fruit salads, compotes, and ambrosias.

Mandarins are an excellent source of vitamin C; one average-size mandarin provides almost 50 percent of the RDA. They are also a good source of potassium and provide useful amounts of vitamin A. An average mandarin contains 37 calories.

Mandarin Calamondin: It is also called Sharon fruit and is often placed in the mandarin orange group because of its loose skin. This citrus item has a thin, smooth rind that easily separates at maturity. Calamondin also has kumquat qualities; it is small and round with loose red-orange skin usually flattened on the top and bottom. Its orange flesh, which is aromatic and very acidic, juicy pulp, and juice, sweetened with sugar, creates a delectable drink. Calamondins make an excellent marmalade. Available December through September.

Mandarin Japanese Orange: More than 100 years have passed since Japanese mandarin oranges were first introduced into the North American continent. Grown on steep hillsides, these oranges are cultivated almost entirely by hand. At harvest time, they are picked by hand, graded according to Japanese standards, and then packed in tissue paper to prevent bruising and to ensure its arrival in good eating condition. The "zipper skin" easy-to-eat mandarin oranges are the most popular fruit among the Japanese people and are very nutritious, with high levels of vitamin C and carotene. Good for eating out-of-hand or for using in fruit salads and desserts. Available in December from Japan.

Mandarin Satsuma: Satsumas are the first mandarin variety of the season. Their most notable quality is that they peel and segment very easily. The exterior will display a light orange, but the interior will be a bright orange. This pebbly textured fruit varies in shape from flat to necked. Satsumas have a mild, sweet flavor, are full of juice, and have virtually no seeds. Available mid-October through December.

TANGELOS

Tangelo: A tangelo is fruit that is a cross between a tangerine and a pummelo. The two most popular varieties, Orlando and Minneola, are available December through April. Other varieties include Nova, Early K, and Sampson.

Tangelo Minneola: Minneolas have a knob-like stem end, a tart, sweet flavor, and few seeds. The Orlando variety contains more seeds than the Minneola and has a mild, sweet flavor. It is available from late November to January. Almost 60 percent of mandarin acreage in California and Arizona is in Minneolas.

Tangelo Orlando: Orlandos are slightly flat in shape and rather large in size. This early producing tangelo is noted for its juicy, mildly sweet flavor. Orlandos have a slightly pebbled texture, good interior and exterior color, many seeds, and a tight fitting rind. Available November through January.

TANGERINE

Tangerine: The tangerine was once called the kid-glove orange because the loose skin came off so easily, without soiling the fingers. It probably originated in Asia, but we first acquired it through the port of Tangier in Morocco: hence, tangerine. The pebbly skinned variety is traditionally seen at Christmas.

Algerian and D'Ancy, the most plentiful varieties, are available November through January. The D'Ancy is the most popular Florida variety and accounts for some acreage in the California desert and coastal interior areas. It is an easy-peeling fruit, deep orange-red to scarlet in color. Harvesting begins in December. Other varieties include Fairchild; Kinnow; Robinson (similar to the D'Ancy); the Clementine, imported from Spain and North Africa; the sweet and nearly seedless Satsumas; the large-seeded Kora; and the Kangpur lime, a tangerine with a sour flavor.

Tangerine Fairchild: Fairchilds are the earliest tangerine variety available. This tangerine displays a good orange exterior color with a bright orange interior. The texture will vary from smooth to somewhat pebbly, and the fruit will generally bear many seeds. Fairchilds are known for their "zipper skin," as they peel easily. Available mid-to-late October through January.

TANGORS

Tangor Temple or Royal: Tangors, which are a cross between a tangerine and an orange, tend to be large and similar in taste to an orange. They are easy to peel, but contain many seeds. Royal Tangors from California and Arizona are available January through April and Temples from Florida, January through May.

Tangor Clementine: According to some citrus specialists, the diminutive clementine is a variety of the common mandarin; others believe it to be an accidental hybrid planted by Father Clement Rodier in 1900 in the garden of his orphanage near Oran, Algeria. Still others believe it to be a hybrid cross between a tangerine and orange. Regardless, this refreshing fruit is deep-orange in color, with a shiny surface, full of juice, and has a tart, sweet, zesty flavored tender melting flesh. Clementines are used in fruit salads, eaten out-of-hand, served with espresso and toasted pecans, or used to garnish poultry, seafood, and salad dishes. They are a good source of vitamin C and offer a fair amount of fiber. Imported from Morocco and Spain, they are available November through February.

P

Papaya: The papaya is called pawpaw by the British, and fruta bomba in Cuba where the word papaya is an obscenity. The plant is a member of the Paw-paw family, native to the Carib-bean and much of South Amer-ica. Cultivated by the Indians long before Columbus arrived, the papaya is a large, slightly ribbed, thin skinned, green, pear-shaped fruit, which is slightly bland in flavor with orange-colored flesh and the texture of a melon.

Before the papaya became popular in the United States, it was known for its special properties as a meat tenderizer and digestant. The fruit, stem, and leaves of the papaya contain the enzyme papain, which breaks down protein in much the same way as pepsin, another animal enzyme. In the areas where grown, people bruise the leaves and wrap them around any meat they want to tenderize. The seeds, which look like shiny black peppercorns, are edible and can be cut in a blender to the size of course ground pepper and added to a salad dressing or used whole as a garnish, like capers.

Today, the Solo variety is grown in Puerto Rico, Florida, Califor-nia, and Hawaii, with Hawaii being the dominant supplier to the markets of this variety to the western United States. The fruit are small, only five to six inches long, and weigh about one pound. Melon-size Mexican papayas weighing ten pounds or more are grown in the American tropics, from Mexico and southward. Prepa-ration includes peeling, halving, removing the large pocket of little black seeds, then slicing. Papaya halves, unpeeled, are filled with seafood salads or served like grapefruit with a spoon. As a dessert item, the halves are often filled with lime sherbet or ice cream. The fruit is rich in vitamin A, C, and potassium, and a one-half pound papaya contains about 60 calories. Available year-round.

Papayas are not to be confused with the pawpaw fruit grown in the eastern part of the United States. They are as different as an apple and an orange.

Pawpaw: This pawpaw should not to be confused with a tropical fruit referred to by the British as pawpaw and known as papaya. Indigenous to North America, it is nicknamed the banana of the north. The reason is that within the pawpaw's greenish-yellow skin is a many-seeded creamy white pulp that has a sweet custard consistency and a strong, nutty, bananalike flavor with a pungent odor. Many people feel that you have to acquire a taste for them. The fruit are borne in clusters of one to six and vary greatly in size and shape, from two to six inches in length, one to three inches in width, and from elongated to rounded. Generally, they are not picked until after the first frost. When first picked, many people wrap them individually in tissue paper and store them in a cool place until soft. Pawpaws are served fresh and eaten like melon or papaya. They are also used in puddings, marmalades, beer, brandy, and custard pies. Pawpaws are high in food value, with more than 430 calories per pound; they contain vitamin A, some of the B vitamins, and traces of minerals. Although not grown commercially, they are available in the eastern United States during the late summer months.

Passion Fruit: It is also called granadilla. Native to Brazil, passion fruit is the edible fruit of the passion flower. The name resulted after early Spanish missionaries saw the passion flower during Easter and Lent, and then named it in honor of the passion of Christ. It is a small, round fruit the size of a large egg, with wrinkled red, yellow, or purple-brown skin, yellow flesh, and many small, black, edible seeds. The pulp has an intense aromatic flavor, while the texture is jelly-like and watery. Passion fruit is ripe when the skin is old looking. Think of passion fruit as a flavoring or sauce. It will be easy to use fresh or cooked. For preparation, slice off the shell tip, then spoon the pulp into a bowl, discarding the rest. It is widely grown in tropical areas for its delicious flavor in beverages, fruit salads, sherbets, jams, and marmalades. A three and one-half ounce serving has 90 calories and contains vitamins A and C, and potassium. Available February to April and August to December from Brazil and other tropical areas.

Besides the purple granadilla, there are sweet granadillas and giant granadillas. The sweet granadilla is oval shaped, three to six inches long, with tough, leathery, orange-brown skin and white, translucent flesh.

The giant granadilla is oblong and measures up to ten inches in length. Inferior to the other two types, it resembles a fat zucchini, with yellow-green skin and a mass of purple, sweet-sour flesh. It is called watermelon in the West Indies.

PEACHES

Peach: Any kind of fruit was a *malum*, (that is, apple) to Romans, so when they saw the first peach imported from Persia they called it *Persicum malum*, Persian apple. Native to China but developed in Persia, the fruit first traveled to Europe before coming to America with early colonists. Peaches are related to the genus of plums, almonds, and cherries. Today they are one of the most popular fruits in America. They are used for breakfast toppings, salads, garnishes for cooked meats, shortcakes, cobblers, pies, puddings, frozen desserts, canning, and for eating out-of-hand. Easily damaged, peaches should be handled very carefully.

Varieties are classified by two methods: by stone tenacity (either clingstone or freestone) and by color (white or yellow). Virtually no clingstones are sold in the supermarkets today, but they are still widely used for commercial and domestic canning. Today, there are several hundred peach varieties grown, with the newer varieties replacing the old favorites such as Elbertas, Hales, and the Rio Oso Gems. They are recommended for low-cholesterol, low-fat, low-sodium-weight reduction, or diabetic diets. Peaches are a good source of vitamin A, calcium, and potassium. A medium-size peach contains about 40 calories. Peaches are available from May through October with the peak season in July and August. Some of the more popular commercial varieties follow:

Peach Dixigem: Freestone, yellow flesh, and good for canning, the dixigem is available from the East Coast in mid-July.

Peach Donut®: This peach is also known as the Chinese flat peach or peento peach. It is a direct descendant of the flat peaches of China. The first tree, however, came from Java to England, where it was grown under the name of Java peach. Imported to America sometime previous to 1828 by William Prince from Long Island, these first trees were lost sometime around 1869, probably to a heavy freeze. The variety reappeared in America when P.J. Berkmans from Georgia brought seeds back from China. The peento was one of them.

With light yellow skin, white flesh, two and one-quarter to three and one-half inches in diameter, and a shape that is flat, round, and drawn in at the center, this peach has an exceptionally sweet, juicy, peach flavor. The Donut® peach can be used the same way as a regular peach: canned, in fruit salads, or eaten fresh out-of-hand. There is only one small seed in the center (freestone). They are a good source of vitamin A, calcium, and potassium. Grown in the heart of Washington State, they are usually harvested in July and August.

Peach Early Red: Freestone, white flesh. Available from the East Coast in early July.

Peach Elberta: Freestone, yellow flesh. Available from Washington in late August.

Peach Fay Elberta: Freestone, yellow flesh. Excellent for canning or freezing. Available from California in late July.

Peach Flavorcrest: Semifreestone, yellow flesh. Available from California in early June.

Peach Hiley: Freestone, white flesh. Available from the East Coast in late June.

Peach J.H. Hale: Freestone, yellow flesh. Available from Washington in late August.

Peach Jerseyland: Freestone, yellow flesh. Available from the East Coast in mid-June.

Peach June Lady: Semifreestone, yellow flesh. Available from California in early June.

Peach O'Henry: Freestone, yellow flesh. Available from California in late July.

Peach Red Haven: Freestone, yellow flesh. Excellent for canning and freezing. Available from Washington in mid-July.

Peach Red Top: Semifreestone, yellow flesh. Available from California in late June.

Peach Springcrest: Semifreestone, yellow flesh. Available from California in early May.

Some more of the eastern varieties include Blake, Red Globe, Coronet, Loring, Junegold, and the Rio Oso. Other western varieties include Gemfree, June Lady, Cal Red, Cassie, and Flamecrest.

PEARS

Pear: The pear is native to Asia and Europe and related to the apple family. The varieties brought to America by the early colonists came from Europe. Pears fall into two different classes. The European is bell-shaped, the flesh is soft and succulent, and the skin is yellow or red. Asian pears are round, with green-yellow or green-russet skin and crunchy flesh. In testing for ripeness, you cannot depend on color. Bartletts will turn from green to yellow. Some red varieties turn from dark red to bright red. Others remain dark red. Winter pears–Anjou, Bosc, Comice, and others–change very little or not at all. The best ripeness test is to gently apply pressure, with the thumb, near the base of the stem. If it yields slightly, it's ripe. The pear season usually begins with early Bartletts in August, then continues through May with the Bosc or D'anjou pear. About 90 percent of the European pears sold in the United States are grown in California, Washington, and Oregon. The typical pear is low in calories–about 100–and contains some B vitamins and minerals. The leading commercial pears grown follow.

Pear Anjou: It is sometimes called D'anjou and is considered a winter pear. A medium to large pear, the Anjou is short stemmed, oval to globular in shape, and has yellow-green skin that may be red-blushed or dull crimson shaded. The flesh is creamy white, juicy, and has a somewhat spicy flavor. The Anjou is excellent for salads and eating fresh. Available from October through May.

Pear Bartlett: It is considered a summer pear and is famous for both its eating and cooking qualities. The pear has a clear yellow skin with red blush when ripe and is extremely juicy, with a smooth and almost buttery texture and very flavorful flesh. Excellent for canning, fruit salads, baked desserts, and eating out-of-hand. Available from August through December.

Pear Bosc: Considered to be a winter pear, the Bosc is a medium to large pear, with a long stem and a long tapered neck, is cinnamon-russet brown in color, and the flesh is creamy. It is valued as the best cooking or baking pear. Available September through May.

Pear Comice: Considered a winter pear, it is a small to large pear, with a short stem and a squatty pear shape. The Comice has greenish-yellow skin with a red blush when ripe, creamy white flesh. It is juicy, sweet and favored for gift boxes or fruit baskets. Available October through March.

Pear Forelle: A golden pear with a bright red blush and a blunt bell shape, this pear has good eating qualities and cans well. Available from October through February.

Pear French Butter: A medium-size pear resembling the green Bartlett pear, it is juicy, with a delicate texture and a slight lemon flavor. It is excellent for fruit salads, fruit desserts, and eating out-of-hand.

Pear Nelis: It is considered a winter pear. The Nelis pear is small to medium size, with a round shape and light green russett skin. It has excellent cooking qualities. Available October through March.

Pear Red Bartlett: It is the same as the regular Bartlett except for its red color that changes to a dark crimson as it ripens. Uses and availability are the same as for the Bartlett.

Pear Seckel: Smallest of all pears, juicy, sweet and often bite-size, it is excellent for eating out-of-hand. Skin can be either green with a dark red blush or nearly full red. Available August through January.

Two other commercial European variety pears are the late-winter Packham pear, a medium to large pear with a lumpy surface; and the Eldorado, a winter pear that resembles the Bartlett.

Pear Asian: Also referred to as Chinese pear, apple pear, oriental pear, nashi pear, Japanese pear, Nijisseiki or 20th Century pear, Kosui pear, Hosui pear, and Kikusui pear, this is the oldest cultivated pear known. Bursting with juice but crisp like an apple, it is a relative newcomer to the American market. The fruit has a clear yellow, brown, or yellow-brown skin and is excellent for eating out-of-hand or for use in salads. Commercial supplies are available from California, Washington, Oregon, and Japan, where they originated.

The most popular variety is the 20th Century, which harvests in September. Asian pears should be selected on aroma rather than firmness. Unlike most pears, this fruit is ripe even when it does not yield to pressure. An average-size Asian pear contains 61 calories, a fair amount of vitamins A and C, and is low in fat. Available July through December.

PERSIMMONS

Persimmon: There are both native-American and Oriental persimmons. All are orange in color except for a small, purple Mexican variety. The two that are native to America are not grown commercially and are confined to Texas and the southern states where they can be found by the roadside, in hedgerows, and in open fields. The fruit is small, dull orange in color, and the flesh is very astringent but edible after a frost. (They are popular with opossums, raccoons, and foxes.)

There are a thousand or so varieties grown in Japan with some grown as ornamentals. The leading Oriental variety grown in this country is the Hachiya; others include Fuyu, Tamopan, Tanehashi, and the Mandarin. Most of the persimmons grown commercially in the United States come from California. The astringent or puckery taste of which they are accused is from misguided individuals who have eaten them while still unripe and have a distorted idea of what this delicious fruit really tastes like. The two most popular varieties in the United States are Hachiya and Fuyu.

Persimmon Hachiya: The most popular of the persimmon family, it is especially sweet and spicy when ripe. It is slightly pointed, with a smooth, bright-orange skin and burnt-orange flesh with a few small, black, edible seeds. Only when the fruit is soft (which indicates ripeness) is it ready to eat. Although the sourest of all fruits when green, it has a delicate flavor and a custard consistency when fully ripe. It is eaten out-of-hand or used in salads, pies, fruit salads, and custards. Available from California from October to December.

Persimmon Fuyu: It is also called marus. This persimmon is somewhat flat with a smooth, brilliant-orange skin and burnt-orange flesh that is firm and can be eaten like an apple. It is mildly sweet with a spicy flavor and is usually eaten hard. It can also be sliced and used in salads or desserts. Seeds, of which there are few, need to be discarded before use. Available from California from October to December.

Persimmon Sharon Fruit: A sweet-tasting, apple-like persimmon imported from Israel, Sharon fruit resembles the Hachiya and Fuyu varieties of persimmon in taste. However, it has a red-orange color and is more round than those persimmons. The fruit is seedless and can be eaten-out-of hand. Available late November through mid-February from Israel.

Physalis: It is also known as ground cherry, husk tomato, goldenberry, cape gooseberry and poha. Native to South America, this richly yellow, opaque fruit is a relative of the Mexican tomatillo, resembles a small cherry tomato in size and shape, and is surrounded by a crisp parchment bladder, twisted closed at the tip like a party favor. It has a thin, waxy skin that surrounds a very juicy, dense pulp of the same color, swirled with soft tiny seeds and an unusual taste that is part tomato, strawberry, gooseberry, and grape, and yet it is sweet, pleasant, and acidic, with a light bitter aftertaste. Remove husks before using or serving. Physalis are used in salads, relish trays, candies, stuffings, pies, and preserves. Available intermittently from New Zealand and California.

PINEAPPLES

Pineapple: The pineapple is native to tropical America and was first encountered by Columbus in 1493. The word comes from the Spanish *piña,* meaning pine cone, and pineapples are still known to Latin Americans as *piña.* Pineapples may not be native to Hawaii, but they were growing there when Captain Cook "discovered" the islands in 1778; pineapples were apparently brought by the Polynesians when they first arrived.

Today, pineapple is one of Hawaii's major crops and is so closely associated with the islands that it is often used as a symbol for Hawaii. But interestingly, the variety of pineapple grown commercially (Smooth Cayenne), was not developed in the islands, but in the hothouses of Europe. At the time of the California Gold Rush, wild-growing Hawaiian pineapples had become well established, and thousands were gathered and shipped to San Francisco where they were sold for fantastic prices.

Captain John Kidwell is credited with starting the pineapple industry in Hawaii. When Kidwell was persuaded to change to sugar cane, he went broke and his good fortune was relegated to history. It was not until James Drummond Dole arrived in 1899 that the pineapple's future was assured. In 1901, he acquired 60 acres on the Wahiawa plateau and organized the Hawaiian Pineapple Company. The Great Depression of the 1930s hit Dole hard, and in 1933 a reorganization was undertaken. Castle and Cooke executives took over management of the Hawaiian Pineapple Company, which eventually became what is now the Dole Food Company.

There are only six varieties of pineapples that are commercially important: Smooth Cayenne, Red Spanish, Queen, Pernombuco, Sugarloaf, and Cabazoni. The Smooth Cayenne is the most important, weighing between three to five and one-half pounds. The flesh is yellow and has a high acid and sugar content. Cold temperatures are harmful, and they are susceptible to bruising. Two slices (about four ounces) of pineapple contain 90 calories and 35 percent of the RDA of Vitamin C. Pineapples are a good source of fiber and potassium. Available year-round.

Pineapple Baby: It is also called the minipineapple. This fruit, first cultivated by the Chinese in the early 1600s, resembles a regular pineapple but is only about five inches tall and four inches in diameter. The baby pineapple has a full pineapple flavor and unlike the regular pineapple, the center core is tender, crisp, and sweet. The Sugarloaf variety is grown to be small and is not an immature pineapple. Baby pineapple contains calcium, iron, potassium, and vitamins A and C. Uses are the same as for regular pineapple. Available at various times from Hawaii.

Pitaya: Widely cultivated in the American tropics, pitaya fruit is ovoid in shape with bright pink or red skin. Sometimes measuring more than three inches long, the fruit has large, leaf-like scales on its surface. The flesh is white, juicy, and filled with several tiny seeds. This type of fruit comes from a climbing cactus that bears night-blooming flowers. Another Mexican wild plant variety produces fruit that is round in shape and approximately two and one half inches in diameter. The skin is covered with small clusters of spines that when brushed off reveal a red, fully ripe fruit. The sweet flesh is dark red to purple. Pitayas can be eaten out-of-hand or used in drinks, sherbets, and preserves. Available from Mexico.

PLUMS

Plum: Plums are native to America, Asia, and Europe. Worldwide, they are the second biggest crop of stone fruits grown, following peaches. Humans have consumed plums since Biblical times. They have good nutritional value and are excellent for pies, salads, desserts, jams, puddings, and eating out-of-hand.

Most of the commercial plums grown in the United States are the red and yellow Japanese varieties. European varieties are the blue and purple fruit. Damson plums, a small-fruited American variety, are considered the finest plums available for jams and preserves. Plums come in an array of flesh and skin colors, with flavors ranging from quite tart to extra sweet. With over 200 varieties grown in the United States, only 20 to 30 varieties are used commercially. Some of the main varieties available from California follow.

Plum Black Beauty: Black plum with yellow-red flesh. Available early June.

Plum Casselman: Dark red plum with deep yellow flesh. Available late July.

Plum El Dorado: Dark red plum with amber flesh. Available late June.

Plum Friar: A large plum with deep black skin and amber flesh. Available late July.

Plum Kelsey: A large green plum that turns yellow-red when ripe with greenish-yellow flesh. Available late July.

Plum Laroda: A large, deep-red plum with yellow flesh. Available July.

Plum President: A large, deep purple plum with light yellow flesh that looks like an oversize Italian prune. Available mid-September.

Plum Queen Ann: Deep-mahogany plum with light yellow flesh. Available mid-July.

Plum Queen Rosa: Red-purple plum with yellow flesh. Available late June.

Plum Red Beauty: Bright red plum with yellow flesh. Available late May.

Plum Roysum: Light red plum splashed with yellow having light yellow flesh. Available mid-September.

Plum Santa Rosa: A purple plum with yellow flesh. Available mid-June. Other varieties of plums include Elephant Heart, Burmosa, Wickson, Simca, Mariposa, Tragedy, and Duarte.

Plum Prune: European plums are always blue or purple and usually sweet. Smaller and less sweet than the Japanese varieties, these are the plums that when dried are sold as prunes. The Italian plum–the main commercial variety–is grown in Washington, Oregon, and Idaho. Another variety grown for prunes in the Midwest to East Coast is the Stanley. Available mid-September.

Plumcot: A hybrid cross between plums and apricots, the plumcot looks like a large, peach-red apricot and combines the flavor of both fruits. Another hybrid of the same fruit is called the pluet. It is deep-purple in color and has the same characteristics as the plumcot, although the texture is somewhat firmer and juicier. They are specialty fruits in very limited supply.

Plum Spanish: Also known as mombin or hog plum, the Spanish plum is native to tropical America. The fruit is round or oval in shape with a light brown or purple skin and is one to two inches in diameter. The flesh is soft and yellow, and when eaten fresh, this plum has an acid, spicy flavor resembling cashews but less aromatic. The fruit is also boiled or dried. Not grown commercially.

Pomegranate: Also called the Chinese apple, the Romans called the pomegranate *pomum granatum*, that is, "apples with many seeds." A symbol of fertility in some cultures, the pomegranate has been used medicinally by herbalists for inflammations such as rheumatism and sore throats. It is also a source of chemicals for tanning leather and treating tapeworm infection, and the flowers are used in making a red dye. Pomegranate juice is a chief ingredient in grenadine syrup, a popular flavoring of wines, cocktails, and confections.

Native to southeastern Europe and the Himalayas, they have been cultivated for centuries in the Near East and North Africa, they eventually migrated to Europe where they became popular in France. Brought to the Americas by European settlers, they are now grown commercially in California. Inside the pomegranate's red or orange leathery skin is a spongy-soft, white membrane that encloses a mass of shiny crimson jewel-like seeds that are edible. The pomegranate has a sweet, aromatic flavor and sometimes weighs up to a pound. Used for garnishing salads and desserts, as well as juice, it can be eaten fresh, usually by sucking the flesh from the seed. An average-size fruit contains about 104 calories. Available August through December from California.

Solomon sang of the pomegranate in the Old Testament, and Theophrastus described it in 300 BC as a valuable fruit.

Q

Quince: Known as the "golden apple" in Greek and Roman civilizations, the quince is related to the pear and a member of the rose family. The ancient Romans believed the power of the quince would ward off the evil eye. Cultivated for thousands of years, it still has a small but loyal following. Introduced to Northeast America in the seventeenth century, quince trees were a common sight in most gardens. Today, the fruit is grown in the Mediterranean, the Middle East, and the United States.

The quince resembles a large pear without the prominent neck, and some varieties are almost round. The fruit turns from green to yellow as it ripens, and the flesh is varying degrees of yellow and is acidic, hard, and rather unpalatable. The quince is superb for making jams, jellies, marmalades, and syrups. Available August through February from California.

Quince Perfumed: Also known as fruiting quince, this apple-like fruit with white flesh is greenish-yellow in color and somewhat wrinkled. It is covered with a light brown felt, is three inches or more in diameter, and has a woody calyx. The large fragrant fruit is inedible when raw, being very astringent and acidic uncooked. Perfumed quinces are used in making jams and jellies. Available from California, October to February.

Quince Pineapple: One of the improved varieties developed by Luther Burbank, the fruit is round or pear shaped with golden-yellow skin and white flesh that has an acidic and somewhat pineapple flavor. It grows about three times the size of an ordinary quince. It is a member of the rose family and is related to apples, pears, plums, and cherries. Pineapple quince must be cooked and is used in making jams, jellies, marmalades, and syrups. Available from California, August to November.

R

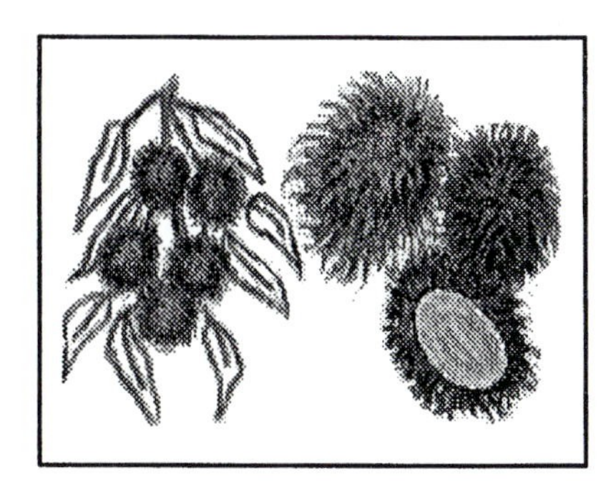

Rambutan: Also called hairy litchi, this fruit is native to Malaysia and is a close relative of the lychee. Grown in clusters of 10 to 12, the fruit is oval, three to four inches in length, crimson red in color, and covered with a mass of soft fleshy spines. The outer skin, thin and leathery, is easily torn off, exposing the translucent juicy flesh that is somewhat acidic like a grape. The fruit is very popular in Malaysia and Vietnam. Very limited quantities are available commercially.

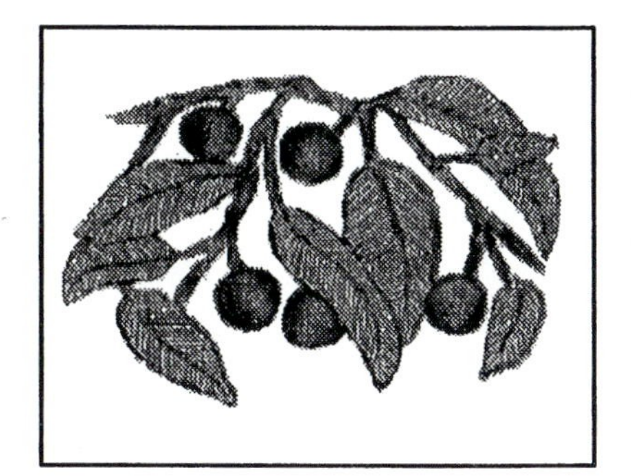

Ramontchi: It is also called Governor's plum. The tree is native to Madagascar. The fruit is small, about one inch in diameter, round, maroon-red in color, and tipped with short radiating styles. The sweet, juicy, edible pulp surrounds several flattened stones and is eaten out-of-hand. Not grown commercially.

S

Santol: Another fruit that is native to the Malaysian region is the santol. While it may not rival the mangosteen or langsat in taste, it is predicted by specialty fruit importers of the United States to become one of the most popular of the tropical fruits. Santols are round to ovate, about two inches in diameter, brownish-yellow in color, with a velvety, thick, tough rind that encloses five segments of white, translucent pulp the flesh contains large seeds. The fruit is generally eaten out-of-hand. Available from the Malaysia area, July through October.

SAPOTES

Sapote Black: Also called chocolate pudding persimmon, the black sapote is native to Central America and Mexico. The black sapote looks much like a fat tomato with a leathery green skin. It has a fibrous, soft, rich, fudge-like flesh that turns from clear bright green to olive tones as it ripens. The ripe pulp can be mixed with sour cream, whipped cream, ice cream, or yogurt for a party dip or dessert. The pulp can also be used as a substitute for chocolate. Available from Florida.

Sapote Mamey: Native to Central America and Mexico, the fruit is football shaped, about six to nine inches in length, weighs one to three pounds, and contains a giant avocado-like seed. The skin is coarse brown and the flesh color ranges from pink salmon to red. Its flesh, when ripe, is creamy, and the flavor suggests sweet potato, avocado, and honey. Rock-hard when picked, the fruit quickly softens to a creamy consistency. To test repeness, mameys should be nicked near the stem to expose the interior; if green shows, the fruit is not ripe and should be ldft at room temperature until it yields to gentle pressure. Their uses include eating from the shell, diced for fruit cups or salads, and for puree. Mameys are a good source of fiber, vitamins A and C, and potassium. They are available July to October from Florida.

Sapote White: It is also known as zapote or zapote blanco and is often called the custard apple because of its smooth, creamy texture. Native to the Central American and Mexican highlands, this fruit has a very sweet, juicy, mild flavor that resembles a blending of fruits. It looks like a misshapen baseball with edible green skin, and the size is comparable to an orange or grapefruit. When mature, the skin has a yellow blush, the flesh can be white or yellow. Ripe sapote should yield to slight pressure when palmed. They can be eaten out-of-hand or added to fruit salads, drinks, desserts, jams, pies, and mousse. Available August through November from California, and April through June from Florida. One-half of a fruit contains about 150 calories.

Sapodillo: It is also called sapota, dilly, wild dilly, nispero, naseberry, and chico. A white latex (chicle) and green fruit that is used in the manufacture of chewing gum is obtained from the tree. Two to four inches in diameter, this round fruit is enclosed by a thin, brown, leathery skin. Its translucent flesh is amber to deep red-brown in color, and the sweet, spicy taste is similar to maple sugar. Sapodillos should be ripened at room temperature, just as avocados and mangoes. They can be eaten out-of-hand, used in fruit salads, or mixed in ices. Available from Florida and Central America.

Sea Grape: It is also called shore grape and sometimes erroneously the raccoon grape. Native to South America, this fruit (pastel green to deep purple) is one-half to three-fourths inches long and bunched in grape-like clusters that ripen late in the fall. Generally found growing on shores and beaches, the fruit makes excellent jelly and wine. Not grown commercially.

Soursop: The soursop is also known as guanabana and is native to tropical America. For centuries, soursop has been used medicinally to treat stomach and intestinal problems. These big, pendulous, heart-shaped, or oblong fruit are deep green in color and covered with many short, fleshy spines. The white flesh is juicy and aromatic, and the texture is somewhat cotton-like, embedded with shiny, black, bean-shaped seeds similar to those of a cherimoya. The pulp is used in making drinks, custard, or chilled sherbet. The soursop has a flavor like pineapple and mangoes. It is closely related to the sugar apple and the custard apple. Available June through December from Florida and Mexico.

Spanish Lime: Native to South America, the Spanish lime is similar to a lychee or longan. This fruit has tough, green-brown skin and milky-colored flesh. It is eaten by removing the skin and sucking the pulp away from the seed. Sometimes available as a specialty item in produce markets.

St. John's Bread: It is also known as locust or carob. The name "locust" comes from the mistaken belief that these pods were the "locust and wild honey" diet of John the Baptist in the desert. They are still used in the Holy Land today as food for horses, cows, and pigs. The tree that produces these pods is native to the Orient and now has migrated throughout the Mediterranean area and Italy (Sicily).

Grafting of the trees is employed to produce a fruit of improved quality, equal to wheat in nutrition. Varieties grown in Sicily are generally preferred for their superior sugar content. St. John's Bread pods are flat, curved, horned shaped, and leathery. They are purplish-brown in color, one inch in diameter, from six to ten inches in length, and contain several pealike, hard, brown seeds in a sweet (50 percent sugar) sticky pulp, sometimes eaten as a natural candy bar. Carob's sweetish pulp tastes like chocolate and is used in flavoring ice cream and in manufacturing syrups and drinks. The powdered pulp is marketed as a substitute for cocoa. Dried carob pods keep well, even without refrigeration. At one time the beans were used as standards of weight for gold and precious gems; hence, the word "carat" is derived from their name. Available in specialty sections of produce departments.

Sugarcane Baton: Originating in India, sugarcane traveled to the West Indies and America and today, grows chiefly in tropical and subtropical areas. It is a source of about half the sugar produced commercially. In Hawaii, where it is called ko, it has become a staple in the people's diet. Cut and packaged in uniform lengths (about eight inches), the stalk has a brown outer bark encasing a fibrous, white flesh that is chewy, somewhat juicy, and sweet. Available year-round from Hawaii.

Sugarcane was a comparatively new plant in the Western world up until the sixteenth century. It was known as far back as the reign of Alexander the Great when one of his soldiers discovered it growing in India. It was on his second voyage from Spain that Christopher Columbus brought cuttings to the Western Hemisphere, planting them in what is now known as the Dominican Republic.

T

Tamarillo: Native to Central and South America, the tamarillo is also known as the tree tomato or tomate dearbol. Available March to November from New Zealand, and October to January from California, this "tree tomato" comes in three varieties. The red variety is very tart and has reddish-golden flesh, while the gold and amber types are mildly tart with golden flesh. The two and one-half to three inch fruit can be used as a fruit or vegetable. People should be careful when cutting tamarillos since the juice stains and is difficult to remove. Do not use wooden cutting boards for the same reason. Tamarillos are not meant to be eaten out-of-hand, they are best cooked. The bitter, meaty flesh has a flavor more like tomatoes than a fruit. The skin should be removed before using. Tamarillos are used as an accompaniment to meat, and in compotes, salsas, relishes, and sauces. A one-half cup serving contains about 50 calories.

Tamarindo: Tamarinds (Tamarindo is the trade name) are believed to have originated in tropical Africa or southern Asia and brought to Europe by Marco Polo. Looking more like a vegetable and resembling brown-shell green beans, this tropical fruit has worldwide recognition as a meat and chutney seasoning. Although tamarinds can be peeled and the pulp eaten raw, the pulp is extracted through boiling for use in preserves, desserts, and sauces. It is also made into a cooling laxative drink. The pulp is brown and studded with seeds, and the acidic flavor is a cross between apricots, dates, and lemons. After the pulp from the seeds is removed, the seeds are cooked and ground into a meal. Both leaves and flowers are eaten in India, and they are also used as corrosives in dyeing. Tamarinds are high in calories, 230 per three and one-half ounces, and contain calcium, phosphorus, potassium, and small amounts of protein, iron, and vitamin C. Available year-round from New Mexico, Florida, Jamaica, Haiti, Grenada, and Mexico.

Tree Tomato: Also known as tomate de arbol (Spanish for "tree tomato"), it is native to Ecuador and Peru. Tree tomatoes are grown on plants that are about nine feet tall and bear fruit that is multicolored, ranging from bright orange to a deep purple. The fruit is not round like a tomato, but oblong and shaped like a football, and the taste is similar to tomato juice, only much sweeter. Used for making drinks, it is also excellent for jams and preserves. Available at times in specialty produce sections.

Wampee: Also called wampi, it is a small tree native to China. The fruits are small, spherical, and about one inch in diameter but with the rough angular shape of an acorn squash. When ripe, the rough, pale-yellow skin splits easily and the tart, white, aromatic, jelly-like pulp is easily removed. The fruit contains three to six small, brown seeds. Wampees are eaten out-of-hand or used in fruit salads, desserts, or fruit juices. Available from Florida, Southeast Asia, and China.

WATERMELONS

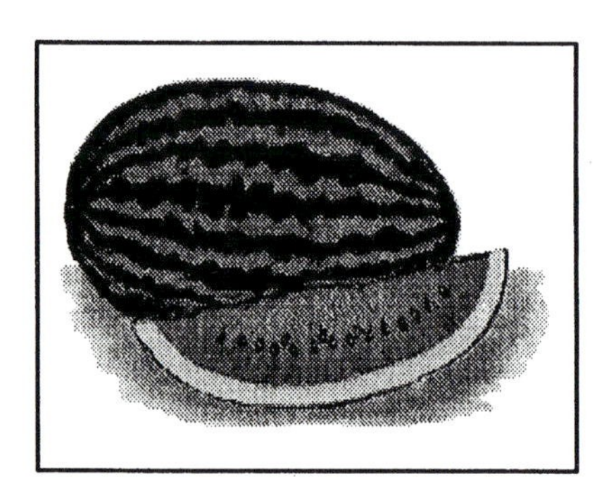

Watermelon: Dr. David Livingston discovered watermelons growing wild in the Kalahari desert in Central Africa. They are now a favorite low-calorie treat throughout North America (Americans consume about 3 billion pounds per year). It is one of the oldest fruits known to man and is now grown on five continents. Watermelon is really a vegetable (*Citrusllus lanatus* of the Cucurbitacae family), cousin to the cucumber and kin to the gourd. They come in many rind colors and with flesh colors of red, pink, yellow, orange, and white. More than 200 varieties are grown in 44 states.

Probably no other question is asked so often by consumers who purchase them than how to select a watermelon that is ripe and flavorful. Ripeness is best determined by the maturity of the stem. A cut stem that pulls off to leave a clean, smooth dish-shaped depression indicates ripeness. The color of the spot where the melon touched the ground is another reliable indicator, since that spot changes from white to yellow as the melon becomes ready to eat. With experience, thumping with the finger can help; the more mature melons give a dead or muffled "ping," and the immature ones produce a higher-pitched sound. If strong enough, holding a melon in the palm of one hand while slapping it with the free hand is a way to select a melon that is not overripe or that has hollow-heart. A good melon will vibrate freely, while the over-ripe and hollow-hearted melon will have virtually no vibration. This method does not eliminate selecting an immature melon.

Watermelons offer a lot of vitamin A, a fair amount of vitamin C, little sodium, and relatively few calories. Ripe watermelons are best kept refrigerated until consumed.

In the winter, watermelons become an off-season specialty, coming largely from Mexico. Florida and Texas produce most of the spring-harvested melons. Summer melons come primarily from Texas, Georgia, California, South Carolina, and Alabama, supplemented by commercial crops in other eastern, midwestern, and western states. Among the leading watermelon varieties are several hybrids and the following open-pollinated ones.

The world record for the largest watermelon is 255 pounds (at the time of this writing), and was grown by Vernon Conrad of Bixby, Oklahoma.

Watermelon Allsweet: Oblong, light green with dark green stripes, 20 to 40 pounds.

Watermelon Black Diamond: Roundish, very dark green, a "cannonball" type weighing 35 to 45 pounds.

Watermelon Charleston Gray: Long shape, light green (grayish) with darker green stripes that are not prominent, 25 to 35 pounds.

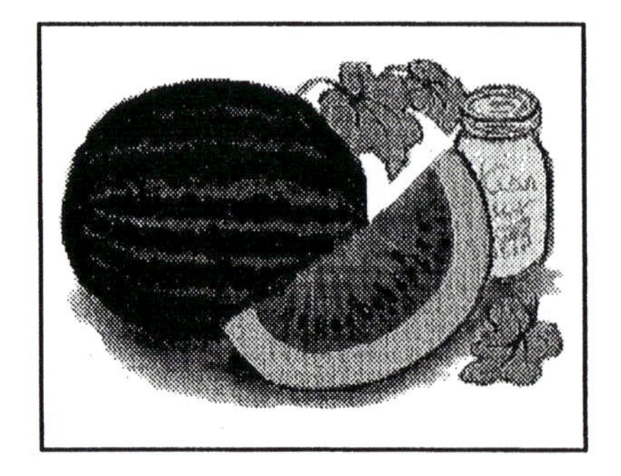

Watermelon Citron: Citron is a preserving watermelon, round, dark green, and striped, weighing about ten pounds, with a thick rind. It is not good for eating fresh but does offer the best qualities for making preserves. The Red Seeded watermelon is the most popular variety of this specialty. (Do not confuse watermelon citron with a citrus fruit that is called citron: a very large, thick-skinned lemon used for candied peel and included with other dried fruit ingredients for fruit cakes in a mixture sometimes called zest.)

Watermelon Crimson Sweet: Developed from the Charleston Gray, oblong and light green with dark green stripes, 15 to 35 pounds.

Watermelon Jubilee: Long shape, light green with prominent dark green stripes, 20 to 45 pounds.

Watermelon Peacock/Klondike: Oblong, light green with dark green stripes, 20 to 35 pounds.

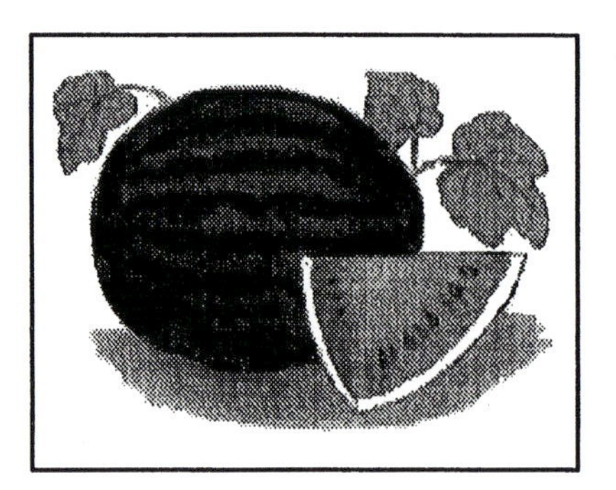

Watermelon Sugar Baby: This variety is roundish and dark green with darker stripes, a small "icebox" variety weighing eight to ten pounds.

There are many other varieties in commercial production. Former favorites, such as Congo, Tom Watson, and Blue Ribbon, may be recalled fondly and inquired about by older consumers, but these are largely replaced by improved varieties with disease resistance and better shipping qualities. New Hampshire Midget was the original icebox type and is still grown in some areas.

Watermelon Yellow Meat: This watermelon is one of the orange-fleshed to bright-yellow fleshed varieties of watermelons found in supermarkets, usually as a novelty, and usually from a local source since they ship poorly. In time, some of these orange and yellow varieties will be commercialized here as they have been in other parts of the world. Some are particularly sweet and flavorful. There is also a yellow-fleshed, small, icebox variety, called Yellow Baby, which is grown in the United States now on a limited scale.

Watermelon Triploid Hybrid: The seedless watermelon is a sweet, crunchy new fruit grown throughout the United States. Developed nearly 50 years ago, the seedless watermelon underwent many years of research to fully perfect its flavor, color, shape, and seedlessness before it was introduced in 1988. The seedless variety is sweeter in flavor, redder in color, crunchier in texture, and smaller and more manageable in size. It is grown in about a dozen states including California, Maryland, Florida, and Texas.

The seedless watermelon is distinguished by its compact (basketball-like) shape, size, and green striping. Each melon weighs about 15 to 16 pounds, fits conveniently in most refrigerators, and when stored properly, remains fresh for extended periods of time.

The seedless watermelon plant is sterile and can only be pollinated by crossing it with pollen from a regular seeded watermelon. In order to accomplish this, growers plant seeded and seedless plants adjacent to one another and pollen-carrying bees pollinate seedless plants to produce seedless watermelon.

The seedless watermelon does contain several white undeveloped seed pods or coats. These soft pods, similar to those in a regular watermelon or a cucumber, are completely natural and quite edible. In fact, nature places these white pods within the melon to hold the fertilized egg and embryo as it develops. Because the seedless melon is sterile, fertilization never occurs so the pod does not harden, preventing it from developing into the familiar hard, black seed. Occasionally, due to the whims of Mother Nature, seedless watermelons will contain a black seed. Watermelon contains no cholesterol, is low in sodium, high in fiber, iron, and potassium, and provides a good source of vitamins A and C. An average slice has only 115 calories.

Vegetables

A

Alfalfa: Also known in China as Mu xu or Muk suk, this plant has been cultivated for centuries as a fodder plant for animals. Alfalfa is a deep-rooted plant that is now being used as a vegetable, especially the young shoots. In the United States, the young shoots are used raw in salads, in stir-fries, and steamed as a vegetable. The leaves are an excellent source of proteins and vitamins. In the Third-World countries, leaves are used in making a concentrate to supplement poor diets. Available year-round.

Amaranth: Amaranth is also known as pigweed. The seeds of this ancient plant were so revered that Montezuma II, the great Aztec king, received 200,000 bushels in annual tribute. Standing shoulder-high and requiring very little care, amaranth produces massive seed heads that may weigh up to eight pounds and contain as many as one-half million seeds. The young leaves are used as a salad ingredient or spinach substitute and are often dried and used as an ingredient for soups and stews. The Indians used to grind the shiny black seeds into meal and use it in porridge or baked in cakes. Available year-round.

Angled Luffa: There are two types of luffa, smooth and angled. Angled luffa is known as Chinese okra (a misnomer), Si gua, and Sze kwa. Smooth luffa is known as sponge gourd, vegetable gourd, loofa, or loofah. Angled luffa is distinguished by its ten sharp ridges and dull green skin. Both types of this long, narrow "squash" are tapered in girth, growing wider away from the stem. The fruit that is on the market, which occasionally may be three feet long, is actually immature. In the tropics it reaches nine feet if allowed to mature. Luffas can only be eaten when young. The older "ripe" fruits are bitter, producing a juice valued medicinally as a purgative. When cooked, this squash has a pleasant earthy flavor and a texture similar to zucchini. Both varieties are used in stir-fries, sauces, and in combination with chicken or seafood, they also complement the flavor of squid. Smooth and angled luffas are eaten throughout the tropical world. Although both mature types can be made into sponges, the smooth luffa are the most popular. In China the sponges are still used in cleaning dishes and scouring pans. They are also used in the manufacture in a range of products including mats and shoes. Available year-round.

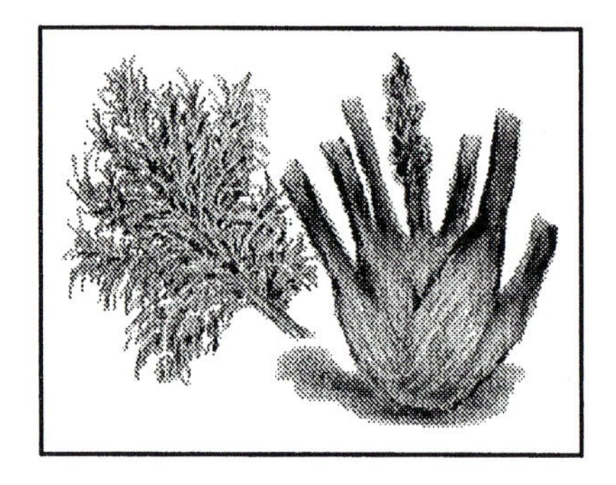

Anise: It is also known as finocchio. Anise should not be confused with the herb anise, even though they both are members of the parsley family. This plant has broad leaf stalks with tops that are lacy–almost like dill–and form a bulb end, often four to six inches in diameter. Having a mild licorice flavor, anise is used traditionally with fish. It is also excellent in lamb dishes, omlets, salads, soups, breads, and can be eaten raw. The stalks can be used in any recipe that calls for celery or creamy celery soup. Available year-round.

Anu: It is also known as anyu, apina-mama, mashua, isanu, and cubio. Anu is native to the high Andes and is a member of the Nasturtium family. Like the primitive potato, the anu only flowers and makes tubers when the day length reaches 12 hours or less. The tubers, depending on variety, range in color from spots of red on the yellowish tuber of the Ken Aslet, to white tubers with a purplish apex of the pilifera (*Tropaeolum pilifera)* variety, to the sparre (*Sylvestre sparre*) variety with slender white tubers. They are generally eaten after being boiled for ten minutes. The flavor ranges from slightly acidic when cooked to slightly peppery when served raw in a salad. In Peru they are eaten half-dried, or boiled and then frozen. The anu is not grown commercially in the United States.

Arrowroot: This waterplant is also known as arrowhead, goo, and Chinese potato. The arrowroot is a bland-flavored root vegetable used in Chinese cooking, and it should not be confused with the arrowroot plant of the West Indies, which is used to make a starch. The starchy corms, like those of water chestnuts, are produced in mud at the end of long stolons. Arrowroot has arrowhead-shaped leaves, three-petal white flowers, and tubers that tend to be egg-shaped (about one and one-half inches long and about an inch wide). The North American species (called wapato or duck potato) was eaten by Native Indians. Arrowroot should be peeled, diced, or sliced before cooking and can be cooked like potatoes or Chinese-style with mixed meat or vegetable dishes. They are usually served on Chinese New Year's Eve, sliced and fried with meat and Hoisin sauce. Available year-round.

ARTICHOKES

Artichoke: Thought at one time to be an aphrodisiac, artichokes were discovered centuries ago by Arabs. Artichokes were first cultivated in Italy during the fifteenth century and then migrated to the Mediterranean areas. Artichokes were introduced to the United States by the French settlers in Louisiana and in California by the early Spanish explorers. Today, California produces almost 100 percent of the commercial crop.

Artichokes are actually the large buds of a thistle plant, and to the uninformed, some instructions in the preparing and eating of them may be necessary. The bud is actually an immature flower head. The tender bases of the bracts (the young flowers and the fleshy base), on which the flowers are borne, provide the edible parts.

The most predominant variety is the Green Globe, deep green in color, and varying in size with a round, but slightly elongated shape. Other varieties, such as the White Globe, Red Dutch, and Giant Bud are not yet involved to any extent in commercial production. Available from California from March to May and October to November.

Artichoke Baby: These are the smaller or primary buds of the artichoke plant. They have no choke (inner part of the regular artichoke that needs to be discarded), and once outside leaves are peeled off, the whole vegetable is edible. Available March through May from California.

Artichoke Chinese: Native to China and Japan, this vegetable is also known as bunge, crosne, Gan lu, or Kon loh and belongs to the same family as mint, sage, and other herbs. This perennial plant produces strings of small, white, almost translucent edible tubers. These tuberous roots are the edible part of the plant and are rarely more than two inches long and three-fourths of an inch in diameter. Chinese artichokes have a delicate, nutty flavor and a texture similar to the water chestnut. They can even be substituted for them. The small tubers are washed and boiled for about five to ten minutes before being served with butter or eaten when nearly cold with a vinaigrette. Other uses include in stir fries, soups, and pickles. Available from September through December.

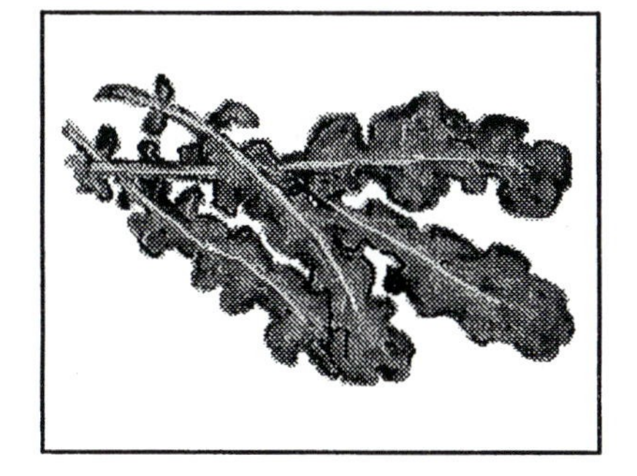

Arugula: Also known as rucola, roquette, rocket, and rugula, it is native to Europe and western Asia. Arugula is leafy green with the appearance of radish leaves, bright green in color, and tender with a slight bitterish, mustard flavor. Arugula is used as a flavoring agent in soups and vegetable dishes as well as imparting an added zest to salads. Some Italian recipes call for ravioli stuffed with arugula, and it is also used in many other pasta dishes. Available year-round from California.

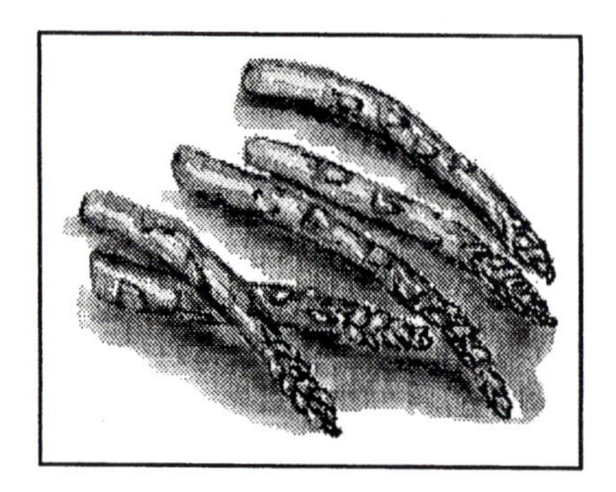

Asparagus: Native to western and central Asia and Europe, asparagus has always been considered a luxury vegetable and was highly prized in ancient Rome, Egypt, and Greece. The spears are shoots from rootstocks called crowns and are cut after only three to seven days of growth. This means that cuttings must be made daily to maximize each shoot at its optimum size. Despite its high cost related to labor-intensive cultivation and seasonality, it is a favorite vegetable for its flavor and a pleasant addition to any meal.

California provides most of the commercial supplies February through April, followed by Michigan and Washington in May and June. There are two general types: white/light green, such as Conover's Colossal and Mammoth White; and dark green, such as Mary or Martha Washington, Viking, and Jersey Hybrids. In Spain, white asparagus is cooked and served cold with salad and vinaigrette dressing, as well as served alone with mayonnaise, and also in omelettes. Dark green varieties are the most popular in the United States. Asparagus is an excellent source of vitamin C and contains a large amount of vitamin A and potassium. A five and one-half ounce serving contains only 20 calories.

The Ancient Greeks believed asparagus was useful in treating toothaches and for preventing bee stings.

B

Baby Vegetables: Baby vegetables are either fully ripe miniature vegetables or immature vegetables picked before fully grown. Baby vegetables are as nutritious as regular-sized vegetables and most offer a tender and more delicate flavor. Currently, there are more than 50 varieties marketed in the United States and Canada. Some of the more popular types are artichokes, avocado (called cocktail), beets, black radishes, bok choy, cauliflower, celery, corn (both white and yellow), eggplant, French green beans, green onions, lettuce, scalloped squash, soft squashes, teardrop tomatoes, and baby zucchini. Most are available year-round from California, Mexico, and Canada.

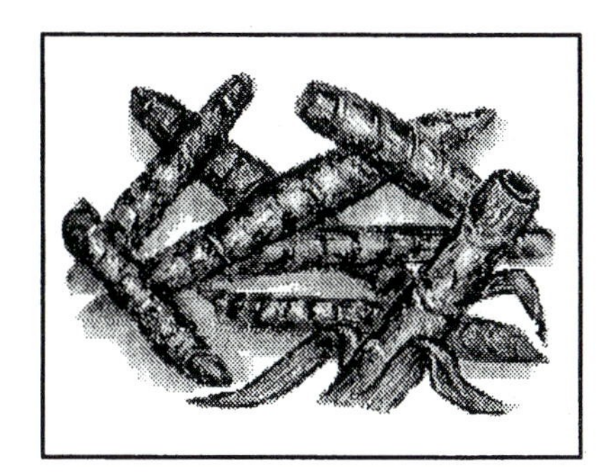

Bamboo Shoot: Known also as Choke-sun, Take-noko, and Chun-sun, these large sprouts come from several species of bamboo, a type of huge grass. Common as an Oriental cooking vegetable, they are relative newcomers to the United States and are usually found in specialty produce sections of grocery supermarkets. Harvested year-round in Taiwan, bamboo shoots are flat, opaque slices from the bamboo plant, which look like asparagus. The texture is tender and crisp if husked and cooked immediately after they are cut. Because they rapidly develop an acrid and bitter flavor after being cut, depending on the species, they usually must be boiled from 10 minutes to 90 minutes before they can be used for cooking. The bamboo shoots for sale in large plastic tubs in Asian markets are not as fresh as some think. They are processed like canned bamboo shoots, only they are sold in bulk. A three and one-half ounce serving contains 27 calories and is a good source of vitamin A, vitamin C, and potassium. Available year-round from Taiwan.

Basella: This member of the Basellaceae family is also known as Ceylon spinach, Malabar spinach, climbing spinach, Luo kui, San choi, or Surinam spinach. Cultivated for centuries in China, it is one of the easiest types of spinach to grow. Basella has been used in China as an important medicinal plant. The cooked leaves and stems are used as a mild laxative and the cooked roots as a treatment for diarrhea. Moreover, the juice is used as a red dye for food coloring, in facial rouge, inks, and printing. A vigorous climber (10 to 20 feet), the stems can become very thick. In preparation, the leaves and stems are cut into reasonable sizes for cooking. Basella is cooked the same as regular spinach. It can be stir-fried or added to soups and stews. Because of its mucilaginous quality, basella is an excellent thickening agent. Available year-round.*

BEANS

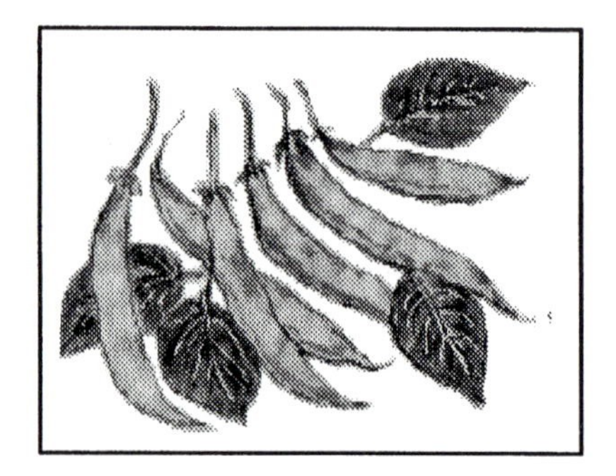

Bean Green: It is also known as French bean, common bean, or kidney bean and is another legacy from the American Indian cultures of Mexico and Central America. The varieties number in the thousands and include colors of white, purple, pink, red, yellow, and variegated. Of the snap or bush beans, only the green and yellow are commercially important. The green bean falls into two categories–bush or vining. Bush beans are grown more widely than

*Part of the confusion in writing this reference guide is the different names applied to the same fruit or vegetable by growers and shippers when it is packed and shipped to the retail market–such is the case with basella. Even in this country there is some confusion created when marketing consultants change the name of a fruit or vegetable in hope that it will generate greater sales.

pole beans. Snap bean pods vary according to pod shape. The three pod shapes most used in marketing are round, oval, and flat. Some of the more popular varieties are Blue Lake, Burpee Stringless, Butterwax (wax), Cherokee (wax), Contender, Puregold (wax), and Kentucky Wonder. Of the edible-pod beans, the snapbeans are the most popular, with the green being the most popular, followed by the yellow wax bean. Most green beans are good sources of vitamin A and potassium and a three and one-half ounce serving contains only 25 calories. Available from October though June with some varieties available year-round.

Ancient Greeks and Romans used beans for balloting in courts. Black beans signified guilt and white beans indicated innocence.

Bean Adzuki: Known as Chinese red bean, Chi dou, or azuki, this bean has always been important in Japan. The pods are short, from two to five inches in length, and contain approximately 12 roundish red beans. The young pods can be used like green beans but are grown mainly for their bean seeds, which are extremely sweet. In China and Japan, they are made into a paste that is used in deserts, pastries, and soft drinks. They make a welcome addition to soups and in salads as a sprouting bean. Available in some Asian markets.

Bean Black-Eyed: Although it is known as black-eyed pea, cowpea, and crowder bean, it is not a pea, but actually a bean. Although black-eyed peas originated in China, it was the Indian and African immigrants that brought this mild-tasting vegetable to the southern United States. It is not a pea, even though it is called one, but actually a bean. This bean is cream-colored and kidney-shaped with a black eye. Use cold, cooked beans in salads, mix cooked black-eyed peas with rice and herbs, or add to soups and stews. Available year-round.

Bean Chinese Long: A favorite in Chinese cooking it is also known as yard-long bean, asparagus bean, snake bean, Chang dou and Dau gok. The long bean is a relative to the black-eyed bean. It is native to southern Asia, but also thrives in Indonesia, China, Africa, some Pacific Islands, and the Caribbean where it is called *bodi* or *boonchi*. Measuring one to three feet in length, the slender vegetable is important in Oriental cooking. Skin color ranges from pale green to dark green, according to variety. The pale-green bean is considered meatier and sweeter than the dark-green bean, which is less delicate in flavor. It is ideal for frying, stewing, and braising. Chinese long beans are low in calories, having about 45 calories per cup, and are a rich source of vitamin A, vitamin C, and potassium, and very low in sodium. Available year-round.

Bean Fava: This unusual bean is also known as English, Windsor, horse bean, and broad bean. Native to the Mediterranean basin and a member of the pea family, it is especially popular in Portuguese cuisine. Larger than even the Fordhook lima bean but more flavorful, fava beans can be served cooked, raw in salads, or with a container of dip. This bean remains a specialty because of its long cleaning process. The tough skin must be peeled and most of the product must be discarded. Fava beans are highly perishable and should be refrigerated immediately. One cup of cooked fava beans contains about 80 calories as well as protein, iron, fiber, vitamins A and C, and potassium. Available from California from April through June.

Bean Garbanzo: It is also known as the chick pea, Indian gram, or ci ci bean. Commonly cultivated in rather hot, dry climates from southern Europe, North Africa, and China, this bean has a nut-like flavor, is round and beige-yellow in color, and has a soft texture. Used in soups, salads, stews, and Middle Eastern cooking. Available year-round from California.

Bean Goa: Another unusual bean additionally known as winged bean, asparagus bean, and Manila bean. Native to New Guinea, it is a tropical legume, highly sensitive to cold, and grown only in the tropics. Since a report to the world in 1975 by the National Academy of Science, more than 70 countries have been introduced to this plant. Almost the entire plant is edible, including shoots, leaves, tubers, pods, flowers, and seeds. It is tasty, nutritious, and high in protein. The pods have four slightly ruffled, equally spaced fins that run its length, so slices resemble a tapered cross. The pods are larger and lighter than the stringbean and when cooked, the flavor is similar to a green bean only meatier, blander, and starchier. Available year-round.

Bean Hyacinth: This attractive plant is also known as Lablab, Egyptian bean, Bonavist bean, Bian dou, and Pin tau. Originating in India, this plant is found throughout the tropics and is now naturalized in the southeastern United States. It is a beautiful ornamental climbing bean with a thick edible root and has edible pods and seeds. In one variety, the pods have a line of rounded tubercles on their edges and are flat and curved. The seeds may be white, black, yellowish, or variously spotted. Hyacinth beans are used both fresh and dried. The mature and dried beans are used as fodder while the immature pods, young seeds, leaves, and tubers are for human consumption. The pods can be cooked in any manner suitable for the green bean and can be used in curries. The mature or dried bean must be cooked well since they contain toxins. Hyacinth beans are usually available July through September.

Bean Lima: Also known as Sieva bean, Madagascar bean, or butter bean, it is native to Central America and northwest South America. The name comes from Lima, Peru, one of the places where European explorers obtained them from Indians. Fresh limas are not usually available in local markets because of the difficulty in shelling them and because the appearance deteriorates rapidly after harvest. Limas are available in two types: butter (small size) and potato (Fordhook or large). They are high in protein and good sources of vitamins A, B1, B6, and C plus fiber and minerals. Their potassium content is good and they are low in sodium. Raw limas can be toxic and should be cooked, and raw lima bean sprouts should never be included in sprouting bean mixtures. Available in dried form year-round.

Bean Potato: It is also known as Indian potato or American ground nut. The Potato bean is native to the eastern half of the United States, from Maine to Florida and west to Colorado and Texas. It has numerous small edible tubers on the roots and racemes of small purplish flowers. The tubers are sweetish and are usually served boiled. The Potato bean was an important food source for the North American Indians and enabled the Pilgrims to survive their first winter in America. Not grown commercially.

Bean Soybean: Known as Eda Mame, vegetable bean, and vegetable soybean, the oil from this bean has been used for centuries as fuel for lamps and stoves in China. Soybeans are thought to have originated in Africa but their development as a cultivated plant took place in Asia where it is a food staple. The soybean's very high protein content has made it a major feed and edible oil crop in the United States. The pods of immature soybeans resemble fuzzy, small peapods and are used in Oriental cooking, steamed or boiled, used like edible pod peas, or cooked in any recipe suitable for fava (broad bean) or lima beans. The curd of soybean is used in making tofu, a high-protein product that blends well with vegetables and sauces. Soybeans are available year-round.

Beet: Native to western Europe and North Africa and cultivated since prehistoric times, the beet was originally grown for its leaves. Early historians tell us that Romans only ate the tops, reserving the roots for medicinal purposes. Root color may be red (most common), orange, yellow, or almost white, but all types taste the same. Beets are grown commercially in 31 states with California, New Jersey, Ohio, and Texas being the main producers. Main commercial varieties include Detroit, Ruby Queen, Crosby, and Early Wonder. Beet roots are low in calories, rich in iron and vitamin A, and used in cooking, sliced or whole (if small), and in vegetable salads. Available year-round.

Beet Baby: It is usually identified as beet greens. Produced year-round, these are available in choices of golden, red, and long red. The goldens are about the size of a large radish with seven inch tops. They have a milder, sweeter flavor than reds, which are more hearty in flavor and have darker tops. Roots and tops should be used together. Beet greens are high in iron and vitamin A, rich in minerals and fiber, and low in calories. Available year-round.

Belgian Endive: Belgian endive is a salad plant, native in wild form to Europe, Asia, and North America. It is also known as wit-loof or chicon. True chicory is often sold as Belgium endive, but this plant is actually endive. True chicory, called witloof or Bel-gium endive in the trade, is sold most often as a tight bundle of colorless leaves forced from the roots of chicory during the winter or enforced darkness. The white head of yellow-tipped, closely wrapped leaves has a mild bitter flavor and is considered a salad delicacy that is sometimes served braised in butter. A three and one-half ounce serving contains 15 calories as well as fiber, iron, and potassium. Usually available year-round with a peak season from November to March.

Bittermelon: It is also known as Chinese bittermelon, Foo gwah, bitter gourd, bitter cucumber, Karella, and Balsam pear, this beautiful climbing gourd is cultivated in Asia for its unusual fruit. They are commonly eaten when very young, but are extremely bitter and usually do not appeal to the Western palate. Bittermelon is a member of the squash family is shaped similar to the cucumber, is about five to ten inches long, and has light green, wrinkled skin. The pulp is spongy with seeds and in shades of white or pink. Immature bittermelons are solid green, while ripe melons are orange. They are popular in India and Ceylon where they are added to curries. They also have some medicinal value since they contain a substance similar in effect to insulin. For the best eating quality, select firm fruit that is green turning yellow. The skin is edible, but the seeds should be removed. Soaking in salt water will remove most of the bitterness. Sometimes served stuffed with meat or seafood often used stir-fried with meat, and it can be steamed, curried, or pickled. Available year-round.

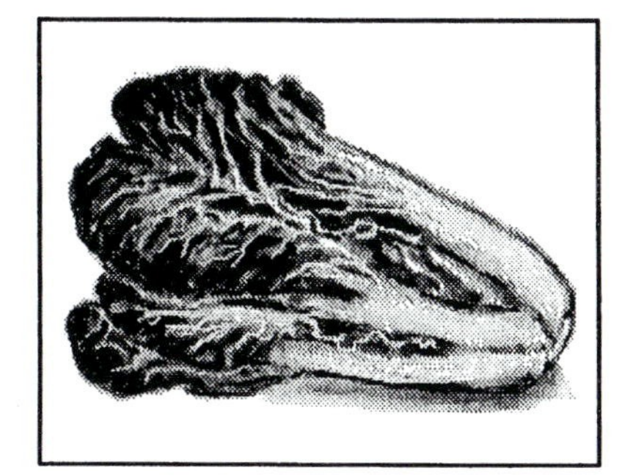

Bok Choy: This popular Asian vegetable is also known as Pak choi, celery mustard, Chinese mustard, Baak choy (Chinese), and Shirona (Japanese). Native to Asia, bok choy is actually an Oriental cabbage and relative to the broccoli, cauliflower, and chard. Resembling white Swiss chard, it has thick white stalks ending in wide, dark green leaves. It is served raw in salads, cooked like spinach, stir-fried with meat or other vegetables, and included in soups. Bok choy contains vitamins A and C, and a one-half cup serving contains about 15 calories. Available year-round.

Boniato: This sweet potato is also known as batata, camote, white sweet potato, batata dulce, and Cuban sweet potato. Native to Central America and now grown worldwide, it is sixth among world food crops. It is a white, creamy textured, tropical sweet potato that tastes similar to the traditional sweet potato, although the flesh is less sweet and cooks up fluffier. Look for white flesh and skins that range in color from pink, purple, cream, to red. Do not base quality on color or smoothness of skin. Boniatos can be baked, boiled, roasted, fried, steamed, sautéed, mashed, pureed, and creamed and used in pies, muffins, puddings, and desserts. Boniatos are high in calories, containing about 115 per one-half cup serving and are a good source of vitamin A. Available year-round.

Broccoli: Native to Europe and the Mediterranean area and a favorite food of the Italians since ancient Rome, its popularity did not initially extend far beyond the Mediterranean. Brought to the United States by the early colonists, this member of the mustard and cabbage family did not gain popularity until the early 1920s, when a group of Italian vegetable farmers in California shipped a trial supply to Boston where it became an instant success. Few vegetables have registered the upswing in popularity in recent years as broccoli.

The head, or edible part of the broccoli, is a dense cluster of flower buds, dark green in color, with an edible stem about six inches long. Broccoli is usually served steamed or boiled as a vegetable, served with a cheese sauce, or broken into pieces for raw salads; often, the floret is served as an hors d'oeuvre with a dip. Like other members of the cabbage family, broccoli is considered helpful in the prevention of certain types of cancer. Rich in vitamins A and C, it is also high in iron, calcium, and potassium and is a good source of fiber. It is low in calories, as a one cup serving contains about 40 calories. Available year-round from California.

The flower buds of the broccoli plant are very attractive and are used sometimes in floral decorations.

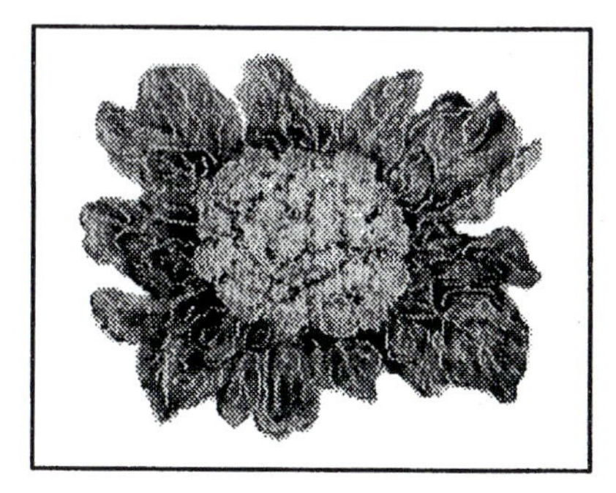

Broccoflower: This is more cauliflower than broccoli. A genetic cross that combines the physical features of cauliflower with the chlorophyll of broccoli. It has a pale green head with a sweeter taste than conventional cauliflower. It is handled and used the same as cauliflower. Available year-round from California.

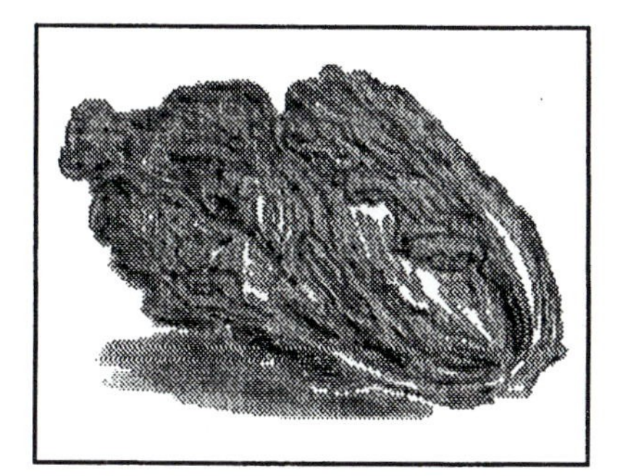

Broccoli Chinese: This member of the Brassica family is also known as Gai laan or Chinese kale. Chinese broccoli is actually Chinese kale that is sold during cooler seasons. It is a light-green vegetable with a long stem, big-flowered leaves, white flowers, and has an earthy broccoli flavor that is slightly bitter. The stems may be peeled if the outer skin becomes tough. Chinese broccoli is prepared and cooked much the same way as Choy sum. It is usually used in stir-fries along with beef, chicken, or prawns. Available December through April.

Brocoletto: It is also known as rapini, broccoli raab, and broccoli rabe. Resembling a thin, leafy broccoli stalk, it is in fact a kind of broccoli, but non-heading, with a pungent-bitter taste. Esteemed by Italians and Chinese for its zesty flavor, it is used in bland foods such as mild potato and pasta dishes. Available year-round.

Brussels Sprout: Native to Europe and a member of the mustard and cabbage family, it is not on record where their place of origin began. First cultivated in the sixteenth or seventeenth centuries, brussel sprouts did not become widely known in this country until the 1920s.

Brussels sprouts are peculiar-looking plants that resemble miniature palm trees with lumps (sprouts) growing on the trunk. The sprouts are borne in the axils where the leaves join the stem. As they begin to crowd the leaf below, sprouts are broken off to create more room and to facilitate harvesting. With the appearance of miniature cabbages and a sweet delicate taste, brussels sprouts can be prepared and promoted as little cabbages and are excellent served au gratin. High in vitamin C, they also contain significant amounts of thiamin, iron, potassium, and phosphorous. Brussels sprouts are considered helpful in the prevention of certain cancers. Available year-round; however, peak season is August through December.

Burdock: Used in China for centuries as a medicinal plant, these long roots are also known as gobo, butterbur, beggar's bottom, and great burdock. Wild burdock grows throughout the United States, Siberia, and Europe as well. Merely a troublesome weed for some gardeners, burdock roots are a thin, long, brown-skinned vegetable staple in Japanese diets. Burdock roots are one to two feet long (they can be longer), about the diameter of a medium carrot, and scruffy brown in color. The flesh is white and somewhat fibrous and quickly becomes brown on contact with air. The earthy flavor is similar to artichoke hearts or salsify. It is used in soups, stews, braises, meat stuffing, stir-fries, and grain dishes. Burdock is also popular pickled; it is often sold wrapped in perilla leaves and is used in the manufacture of soft drinks. Available year-round from California, Hawaii, and Japan.

C

CABBAGES

Cabbage: Native to northern Europe, the wild cabbage originally was loose leafed like collards. Used by humans since prehistoric times when they subsisted on wild plants, cabbages were among the first plants to be cultivated. Related to broccoli, brussels sprouts, cauliflower, collards, kale, and kohlrabi, the common cabbage is a compact head of leaves, green on the outside and whitening toward the interior. It was one of the first European vegetables brought to the New World by the colonists.

Cabbages are classified by leaf type and color. Smooth-leaf varieties such as the Danish are used in cooking and for slaws. Round or with a drumhead shape, it is usually available from July through March and is classified as winter cabbage.

Romans revered cabbage as a medicine. Poultices made from red cabbage were recommended for treatment of warts, tumors, and sores and to prevent rheumatism.

Cabbage Savoy: The dark green, crimped-leaf varieties, known as savoy types, are also used in cooking and slaws. Round-headed with yellow-green leaves, savoy cabbages make an excellent table centerpiece.

Other domestic types are usually pointed or drum-shaped with color variations ranging from light green to black (actually a very dark green). The head is less compact, and the leaves are more tender. These types are usually classified as a summer variety.

Cabbage Red: The red or purple cabbage is essential for the color variation in slaws and salads. Heads are round and solid, and generally more mild and sweet flavored. Red cabbage is popular for pickling as well as salads and cole slaw. Cabbage is readily available year-round and benefits from refrigeration. Cabbage is high in vitamin C, matching that of orange juice, pound for pound. It also offers a wide range of other vitamins and minerals and good roughage.

Cactus Pad: It is also known as nopales, cactus leaves, and nopalitos. Native to Mexico, these are the thorny pads from the cactus of the Opuntia cactus family, the same cactus variety that offers the fruit known as cactus pear. These thorny, green pads taste like tangy green beans. Once the thorns are removed, they can be sliced or chopped for salads or cooked like green beans. Cactus pads are an excellent source of vitamins A and C and contain a fair amount of B vitamins and iron. Available from California, February through November.

Cardoon: Looking like a giant artichoke plant with small, prickly flower heads, this plant is also known as cardoni and cardi. Native to Europe and the Mediterranean area, it is now grown in Australia and Argentina where its popularity has spread like a weed. It was once eaten by pregnant women in the belief it would ensure a male offspring. A member of the thistle family and a close relative to the artichoke, it looks somewhat like celery and has an artichoke-like flavor. Cardoon is used to flavor soups or stews, used in salads, can be deep fried or marinated, and is often served as a separate vegetable. The dried flowers can be used as a substitute for rennet. It is very low in calories and an excellent source of potassium, calcium, and iron, but it is high in sodium at 300 milligrams per cup. Available from California, October through May.

Carrot: Native to Afghanistan, carrots, like celery, belong to the parsley family. They were cultivated originally for medicinal purposes. As their use spread westward, they were introduced to England in the fifteenth century. At that time, stylish ladies of the courts used the feathery leaves to adorn their hair and hats. Nowadays, California is the largest producer of carrots followed by Texas and Canada.

There are many types of carrots. Root colors range from red and orange to yellow and white and vary in shape from round and blunt to long fingered, and pointed. The versatile carrot is used in stews, soups, raw snacks, in mixed vegetables, juicing, cakes, and salads. It is exceptionally high in vitamin A. One average carrot, about two and one-half ounces, provides more than three times the RDA of vitamin A and contains only 40 calories. It is a good source of potassium and crude fiber, and is low in sodium. Available year-round from California.

Carrot Mini: It is also known as the French carrot. This small carrot is one to two inches long and very sweet. It is used for stews, roasts, and relish trays. Available year-round from California.

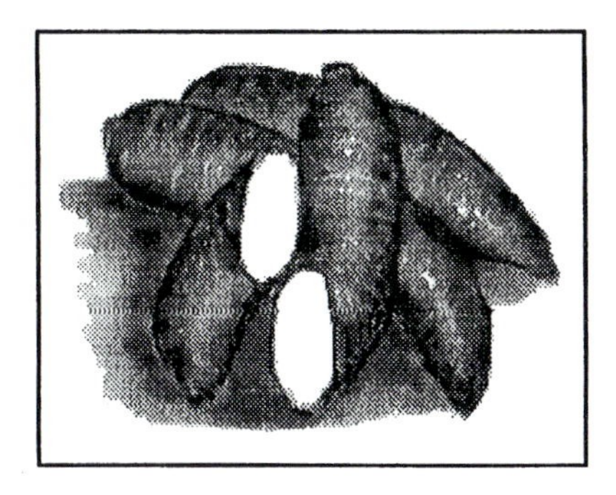

Cassava: Cassava is known as manioc, manihot, mandioca, tapioca, eddoes, and yucca. Because the word *manioc* is thought to have originated with the Tupian natives of Brazil, cassava is thought to be native to Brazil. They were taken to Africa around 1600 by the Portuguese, where the roots were quickly absorbed into the African diets. The cassava leaves are sometimes used by natives as medication for lacerations and treatment of chicken pox, but for millions in Africa, the leaves as well as the roots are used for staple food. In fact, a delicious stew–*ngunza* in Central Africa–is made from cassava leaves. Related to the castor bean and para rubber tree, the cassava's large, two- to eight-inch diameter roots may reach a length of three feet or more and weigh more than 25 pounds.

There are two types of cassava, bitter and sweet, and both are important sources of starch. Sweet cassava is usually used for the table, like potatoes, either boiled or roasted, and has a flavor similar to chestnuts. Bitter cassava, however, contains prussic acid and can only be used after special treatment to remove the poisonous acid. The common tapioca, highly esteemed for making puddings, is derived from the bitter cassava. Cassavas and leaves are available year-round.

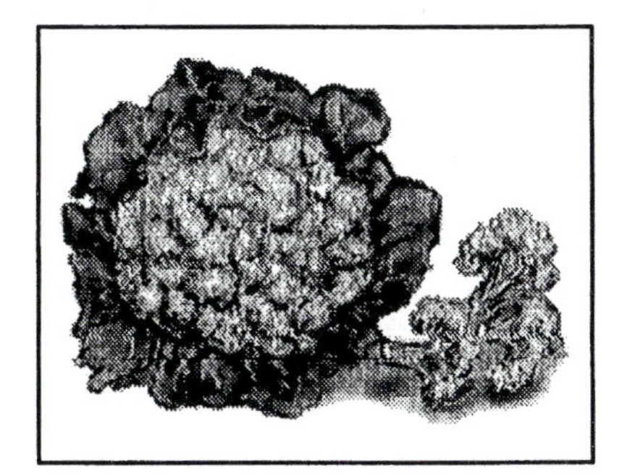

Cauliflower: Native to Europe, it is another member of the mustard and cabbage family and has been cultivated in Asia Minor and the Mediterranean area for more than 2,000 years. While known for centuries in Europe, this vegetable has only been an important crop in the United States since the 1920s. Mark Twain once described cauliflower as "cabbage with a college education." While its first cousin broccoli sends up a stack of flower buds, cauliflower retains its potential flowers in a compact head called a curd. Varieties include those with white, green, or purple curds, with white being the only one grown commercially.

Cauliflower is used as a raw ingredient in salads, broken in small pieces and served raw with dips, used in stir-fries, and often served as a separate vegetable. Like other members of the cabbage family, cauliflower is considered helpful in the prevention of certain types of cancer. It is a good source of potassium and vitamin C, and one cup of cooked cauliflower contains about 28 calories. Available year-round from California, Arizona, Texas, and New York.

Cauliflower Romanesco: The romanesco is a decorative cauliflower, yellow-green in color, and made up of tightly packed small "turrets." It cooks quickly and has a mild taste. Uses are the same as for regular cauliflower. Available September to December.

Celeriac: This rootstalk is also known as celery root, knob celery, German celery, turnip-rooted celery, and apio. Celeriac is believed to have originated in Europe in the Mediterranean area. Grown for its large rootstock, ranging from apple size to cantaloupe size, the lopsided bulb is covered with channels, whorls, crevices, and rootlets. Once cleaned, the light-brown skin and rough exterior can be peeled, exposing a creamy white flesh with a mild flavor and crunchy texture. Celeriac must be peeled and it can be eaten raw, boiled as a vegetable, or used in soups and stews. It is high in phosphorous and a good source of potassium. A three and one-half ounce serving contains about 40 calories. Available year-round.

Celery: Native to Europe and the southern Scandinavian countries, celery, a member of the carrot family, has been known for thousands of years. Originally a bitter, wild marsh plant that was used by the ancient Greeks and Romans for medicinal purposes (to purify the blood), it is now used as an appetizer and cooking ingredient. There are two distinct types that are classified by color: green or golden (white). The green type is referred to as Pascal and comprises the bulk of production. Celery is used raw as a snack, in soups or salads, as a poultry stuffing ingredient, in stir-fries, and as a vegetable served cream-style. It is low in calories, high in fiber content, moderate in minerals, and a fair source of vitamins A and B. Available year-round.

Celery Chinese: Also known as Qin cai, K'an tsoi or Seri-na, Chinese celery is probably a wild form of Asian celery. With thin, hollow stems, it is used as a flavoring herb and vegetable in most of China. The plants, approximately 15 inches high, have leaves ranging from dark green to yellow depending on variety and have a very hardy celery flavor. When used in flavoring, leaves and stems can be used whole or they can be finely chopped and combined with meat and fish dishes. As a vegetable, it is usually used as a stir-fry ingredient. Chinese celery is usually available through Asian markets.

Celtuce: This crunchy vegetable is also known as stem lettuce, asparagus lettuce, Wo sun, Woo chu, and Chinese lettuce. Celtuce is native to China and is now grown commercially in the United States. It is not a cross between celery and lettuce, but a very old vegetable from China. It has a thick, edible stem and romaine-like leaves, which are six to eight inches in length, and can be cooked like broccoli. Young leaves are used in salads, and young stalks are used like celery. Celtuce tastes like a cross between a mild summer squash and an artichoke. Nutritional values are comparable to celery. Available year-round.

Chard: This member of the beet family is also known as Swiss chard, leaf beet, white beet, spinach beet, and silver beet. Chard is believed to have originated in the Mediterranean area and the Near East. No one seems to know why it is called Swiss chard. A member of the beet family, it grows up instead of down and produces large leaves and fleshy stalks, instead of a bulbous root. The stalks may be white or red, depending on variety, the leaf is dark green, and the flavor is mild, yet earthy with a slight bitter undertone. Used raw in moderation for salads, it is also used in stuffings, savory custards, egg dishes, and boiled as a vegetable. Swiss chard is an excellent source of vitamin A, potassium, and iron. A one-cup serving contains about 26 calories. Available year-round from California.

Chicory, Green Heading: Green heading chicory include several plants including those known as endive, escarole, and radicchio, although radicchio is also known as red chicory. A salad plant cultivated from wild forms found throughout most of the Northern Hemisphere, green heading chicory has a loose, pale-green head ranging from six inches to one foot long in length and two to six inches in diameter. It is used in salads or as a cooking ingredient. Available year-round.

Chicory, Green Loose-leaf: Also known as radichetta and green chicory, green loose-leaf is the largest group in the chicory family. Some varieties have thick, curly leaves, while others have thin, flat ones, and the leaves can be indented or smooth. It is prepared and used much the same as the green heading chicory. Available year-round.

Chinese Leaves: This member of the Brassica family is also known as celery cabbage, Pe tsai, or Peking cabbage. This mature plant is generally known as Chinese cabbage. The plants are like romaine, but with a crisp, watery texture and mustard-like taste. There are both round and elongated varieties with solid hearts and open leafy types with broad stalks. Chinese leaves may be eaten raw or cooked. Available year-round.

Chinese Clover: It is also known as bur clover, toothed bur clover, hairy medick, and Cai mu xu. Introduced to Hong Kong and Taiwan during the Communist takeover in 1945, it is now a popular crop in China. A small creeping plant about 12 inches tall with stems containing three to four slightly oval leaves. For best flavor, the plant should be harvested when young. The leaves, leaf stalks, and leaf tips are all usable. The chinese use them in soups, boiled, steamed, and stir-fried. Available year-round.

Chive: Growing in clumps, it is a dainty onion that grows wild throughout Europe and the northern latitudes of the United States. Often sold potted by commercial growers, it is used much like an herb. Finely chopped, they add a delicious flavor to salads, omelets, certain cream sauces, baked potatoes, and cheese dishes. Another variety grown in China, aptly called Chinese chives or Gau tsoi, is very similar in appearance and is used as a seasoning herb also. Available year-round.

Choysum: Actually the flowering shoots of Chinese chard or bok choy, choysum is used like bok choy but is more tender. This is one of the many Chinese vegetables often called a different name for the same vegetable or different vegetables with the same name. The key is to remember that in Cantonese *choy* means vegetable and *sum* means heart or flowering stem. The flowering shoots are the main parts used, with the young, tender shoots in salads. Available year-round.

Cilantro: This very popular herb is also called Mexican parsley, coriander, Pak chee, Yee sai, and Chinese parsley. A favorite ingredient in Mexican, Chinese, and other Asian cooking, it is similar in appearance to the Mediterranean parsley only with broader leaf tops, the herb has a pungent, musty, spicy, and aromatic flavor. It is used in soups, stews, meats, cheeses, pickles, stir-fries, salads, tacos, salsas, and as a garnish. Available year-round from California.

Cipolline: Imported from Morocco, this plant belongs to the shallot, onion, and garlic family. It has a small, round, onion shape, and is slightly bitter. It is used primarily for seasoning. Available as an import.

Collard: This popular southern vegetable is also known as collard greens, coleworts, and collie. Originating in the Mediterranean basin, they were a favorite food of the ancient Greeks and Romans. Collards are nonheading cabbages similar to the wild cabbages from which the heading varieties were developed. Collards have large, thick, tender leaves, medium to dark green in color, with an earthy cabbage-kale flavor. Usually used as a separate vegetable, they can be simmered like cabbage or used in combination with other foods. Collards are low in calories and an excellent source of folic acid and vitamin A. They are also a source of iron, zinc, potassium, and calcium. Available year-round.

CORN

Corn: Also known as maize (the proper word), it is native to North and Central America and was introduced to Europe by the early European explorers and settlers. Strictly speaking, the word "corn" means any particle of grain or any small pellet of anything. Americans adopted the word to describe the grain of a special grass that grew to a huge size with seeds in rows on thick cobs. North American tribes, Aztecs, Mayans, and Incas cultivated corn (maize) thousands of years before the Pilgrims arrived. Aztec and Mayan civilizations were built on a corn-based economy, with corn providing food, currency, fuel, fodder for animals, sugar, and fermented beverages. The sweet or sugar corn we enjoy today is a mutation of Indian field corn.

Cuitlacoche (pronounced HWEET-la-cochay) is a fragrant and spicy fungus cultivated on ears of corn during the rainy season. Formerly called corn smut, it is now a gourmet delicacy. It has a slight smoky aroma and smoky mushroom flavor with the corn flavor dominating.

Corn Sweet: Sweet corn first appeared in the mid-1800s, but major advancements in its evolution did not occur until the 1920s. Today, sweet corn varieties are classified by kernel color: yellow, white, golden, and newer hybrid bicolor types. Corn is served on the cob (usually boiled in water, roasted, or barbecued), in combination with other vegetables as well as in salads, soups, relishes, chowders, casseroles, puddings, and fritters. Corn is low in sodium and supplies a substantial amount of vitamins A and C and carbohydrates. One ear contains about 90 calories. Available year-round.

Corn Ornamental: Also known as zea it is a variety of corn (maize) cultivated for its ornamental foliage or brightly hued grain. The colorfully seeded ears are dried and used as winter decorations. Available October through March.

Corn Strawberry Popcorn: These shiny, little, two-inch ears of corn are a rich mahogany color with bright husks. When picked and dried, they look like huge strawberries in centerpieces and fall arrangements. They are excellent for popcorn. Just store them in a refrigerator for a short time to restore moisture prior to popping. Available during the late fall months.

Cress: It is also known as land cress, garden cress, and English cress. Do not confuse with watercress. This delicate cousin of watercress is a member of the mustard family and grows on soil. Used since the Middle Ages as a seasoning vegetable, it is cooked with other vegetables; sometimes, it is cooked and served separately. Delectable and nutritious, it is sometimes underrated. Available year-round.

CUCUMBERS

Cucumber: Native to India and southern Asia, cucumbers come in a fantastic assortment of sizes, colors, and shapes. Because cucumbers are about 96 percent water and their skin holds in moisture like a jug, they have been used to quench thirst since ancient times. Cultivated in India more than 3,000 years ago, they were favored by desert inhabitants for their cool, refreshing taste.

Cucumbers are divided into three classes: the slicing or table cucumber, the pickling varieties, and the greenhouse varieties. The more popular varieties of table cucumbers are Ashley, Marketer, Palomar, Poinsett, and Straight Eight. Used in salads, for garnishes, and in relish trays, they are also good cooked like a vegetable or steamed and served with a cream sauce. Raw, cucumbers are a good source of vitamin A, and a medium cucumber (10 to 12 ounces) contains about 45 calories. Beyond the traditional cucumber, there are several special varieties available. They include the following.

Cucumber Armenian: This cucumber with a mellow-sweet taste, is 10 to 18 inches long and 1 to 2 inches in diameter. It has a coiled shape with ridges, edible skin and is used in salads, relish trays, and as a cooked as a vegetable. Available from July through August.

Cucumber English: Sometimes called hothouse, European, or burpless, it is an elongated cucumber, 8 to 16 inches in length, 1 to 2 inches in diameter. It is seedless and has a mild flavor with no bitterness. Used the same as the traditional cucumber. Available year-round.

Cucumber Japanese: It is part of the mild-flavored Armenian cucumber family. Similar in shape to the European, this vegetable is nine to ten inches long, has a prickly skin, and is usually smaller than the regular cucumber. Often used in Oriental cooking, it is also used in salads, sliced and served with a dip, or in sandwiches. Available year-round.

Cucumber Lemon: A burpless cucumber with a milder flavor than green-skinned varieties. About the size of a tennis ball with an oval shape, it turns from lemon to golden-yellow as it matures. It is used the same as the regular green-skinned cucumbers. Available from California late May through mid-July.

Cucumber Pickling: A small (two to four inch), short, hard, green, and spiny cucumber with a tart flavor that is especially good for making sweet or dill pickles. Available July through August.

D

Daikon: It is also known as Chinese radish, Japanese radish, Loh baak, and Lo pak, and thought to be native to China. Botanically, these are members of the same group as the small red radish (the Crucifer or Mustard family). While the daikon will grow to 50 pounds or more, the marketable ones average between one-half and two pounds. The most common one available in the North American market is a sweet, juicy, pearly white variety that is about a foot long, round, and tapered at the tip like a carrot. Other varieties are more irregular and greenish, although it is possible to encounter rose, black, and other shades. Daikon can be pickled, stir-fried, or slivered, diced or sliced to add spice to salads, or used in relishes or with seafood. Grated daikon is a traditional accompaniment to Japanese raw fish dishes, such as sushimi. Daikon is very low in calories, having about ten calories per one-half cup. Available year-round from California.

Dandelion Green: Still considered a weed in some parts of the United States, it was once available only to those who gathered them. Only the young plants are used before the blossom stems appear since mature plants become very bitter. The longer-leafed cultivated dandelion with its saw-toothed, barb-shaped leaves are milder than their wild forebear. Used raw in salads or cooked much like spinach and served in cream or cheese sauces, dandelion greens are very high in vitamin A, iron, and potassium and a fair source of other vitamins and minerals. They contain about 35 calories per cup cooked. Dandelion roots contain a wealth of nutrients. It is said that ancient people were able to subsist completely from the dandelion when adverse weather conditions destroyed many of their crops. Available year-round.

A type of beer is made from the leaves and a wine from the crushed flowers of the dandelion. The roots can be roasted or dried and ground to make a beverage similar to coffee.

Dill: Also known as dill weed, it is native to Europe and the Mediterranean area. At one time, it was used as an ingredient in witch's brew. This aromatic herb belongs to the carrot family *Umbelliferae.* The feathery, bluish-green stems are harvested when the stems are about one foot long. Its anise-parsley-celery flavor goes well with fish, vegetables, soups, and salads. Crown dill, a stronger-tasting plant, is used in making pickles, while baby dill is used primarily as a seasoning. Available August through October.

E

Eggplant: This plant is also known as aubergine, a member of the nightshade family, along with tomatoes, potatoes, and some poisonous plants. Originally an Oriental ornamental plant, eggplant got its name from yellow and white fruited varieties with egg-size fruits. During the sixteenth century, northern Europeans called them "mad apples" in the belief that consumption would cause insanity. They were brought to America by Spaniards as *berengenas*, meaning apples of love. Ladies of China once made black dye from dark eggplant skins and used it to stain their teeth black, a fashion at the time.

Eggplant may be white, purple, purple-black, green, orange, or even striped. Under the skin, however, most varieties are pretty much alike. The exceptions are tiny green Thai eggplants, which add crunch to certain Asian dishes, and a bitter orange Thai eggplant used in sweet-sour dishes.

Eggplant is used as a meat substitute, baked, fried, and as parmigiana. Often it is stuffed or broiled. Eggplant is not especially rich in any one vitamin or mineral. One cup of cooked eggplant contains about 38 calories. Eggplant is low in calories, low in sodium, and a fair source of potassium. Large, pear-shaped purple varieties are the most popular in the United States. The long Oriental varieties are slender, curved, have smoother flesh, thinner skin, and fewer and smaller seeds.

Some of the more available commercial varieties are: Black Beauty; Italian; Casper; French Bonde de Valence; Chinese White, Asian Bitter Orange; Rayada; Thai Round Green and White; Thai Round Purple; Italian-Rosa Blanco, Green Egg Shape, White Egg, and Easter Egg; Thai-Green Streaked, Thai Turtle Egg, Applegreen, and Thai Bunch Green; and Pea. Available year-round.

Endive: Native to Europe and the Mediterranean area, endive, escarole, and chicory commonly are confused with each other. In some areas of the country, the names are used interchangeably, adding to the confusion. Endive grows in bunched heads with narrow, ragged-edged leaves that curl at the ends. The center is yellow-white. The taste at the center is mild while the outer green leaves tend to be bitter. Endive is a popular salad ingredient that adds texture and taste. Endive is a good source of potassium and vitamin A. A three and one-half ounce serving contains about 40 calories. Available year-round.

Escarole: Escarole is believed to be native to India. Often confused with endive or chicory, escarole leaves are not at all like the sawtooth-edged leaves of chicory or endive, but are broad and curly. The uses and nutritional values are comparable to those described under chicory. Available year-round.

F

Fiddlehead Green: Native to the northern hemisphere and harvested wild in northern New England and eastern Canada, it is the young shoot of a fern, shaped like a question mark, and has a taste similar to asparagus. The dark-green fiddleheads should be washed first to remove the characteristic brown scales, then it can be steamed, blanched, or sautéed like asparagus and served hot or cold. They are a good source of vitamin A. Available from Oregon, April and May, and from Canada in June.

Frisee: Also known as French endive, this salad green has short, curly, white inner leaves and greener outer leaves. Taste is light with some bitterness in the outer leaves. Uses are comparable to those of regular endive or chicory. Available year-round.

Fuzzy Melon: These pleasant tasting melons are also known as hairy melon, fuzzy or hairy gourd, and Mo gwa. A close relative of the large winter melon, all fuzzy melons are varieties of a species called winter gourds, a member of the Concurbita family, and are thought to be native to Java, where they can still be found growing wild. Fuzzy melons come in two shapes. One is squash-like, narrow, cylindrical, and sometimes crooked; the other looks like a stubby pill capsule. Harvested young, between 12 to 16 ounces, they are easy to spot because they are truly hairy, a condition they will outgrow if allowed to mature. Fuzzy melons are used in a variety of ways: steamed, stuffed, shredded, blanched in oil and then stir-fried with dried shrimp, or cubed and cooked in soups. Available during the early U.S. summer months.

G

Galangal: This fiery tuber is also known as galangal root, Loas, Thai ginger, Java root, Ka or Kha (Thai), or Languas. Galangal, a relative of ginger, was popular throughout Europe in the Middle Ages when it was dried and ground. It resembles ginger, but the rhizomes are larger and pale yellow with zebra-like markings and pink shoots. It is not to be eaten raw, as those who have bitten into it can attest. It has a fiery medicinal taste and the texture of a wood chip. Pounded in a mortar with seasonings such as lemon grass, chili peppers, shallots, or garlic, galangal is an important ingredient in Thai curry pastes. Its ability to curb nausea and settle the stomach is supposedly as great as ginger. Lack of commercial growers and scarcity in the market contribute to the high cost of galangal. Available from the Fiji Islands.

Garlic: Native to Central Asia, it has been cultivated for thousands of years. Although wild garlic grows here, the cultivated varieties came to the United States from Europe. Garlic is a member of the lily family and a close relative of the onion. The bulb is the plant part that has been revered in songs and stories and treasured for centuries. A single bulb is composed of 8 to 12 sections called cloves, which are held together by a parchment-like covering. Garlic contains the amino acid alliin, which scientists say has antibiotic and bactericidal effects. It has been long accepted as a purgative, and is believed to promote cardiovascular activity and have a beneficial, soothing action on the respiratory system.

Garlic supposedly works against food poisoning. Microbiologist David Hill of the University of Wolverhampton said the oil in garlic works against bacteria that cause food poisoning in the digestive system, but only if it is taken regularly. He monitored the total normal population of bacteria naturally present in the intestines and found that in the presence of garlic, normal bacteria grew, but pathogens were eliminated. Researchers think that sulfur compounds found in garlic may be the specific ingredient that fights food poisoning.

Garlic (white or purple-skinned) is a pungent member of the onion family and is mainly used as a seasoning for foods. It is used in sauces, soups, breads, meats, salads and many Italian dishes. Heat and handling affects garlic's potency. Raw garlic is stronger in flavor than cooked garlic. The longer it cooks, the more delicate the flavor. Roasted, it becomes sweet and is often used as a spread on bread. It is also used in a lot of pickling recipes. Whole cloves and large pieces give off a gentler flavor than cut or minced. Garlic is about 75 percent water and contains about four calories per clove.

To remove the garlic scent from hands, rub then on stainless steel (your kitchen faucet, for instance). Available year-round.

Ancient Egyptians required their slaves to eat garlic, believing it gave them strength, and Romans believed it gave soldiers courage. Gladiators were instructed to eat garlic to make them capable of greater feats of strength in the stadium.

Garlic Elephant: It is similar to regular garlic in appearance, but has a milder flavor and is much larger. Heads can weigh as much as 12 ounces and be six inches in diameter, and one clove can be as big as an entire head of regular garlic. Its uses are comparable to regular garlic. Available year-round.

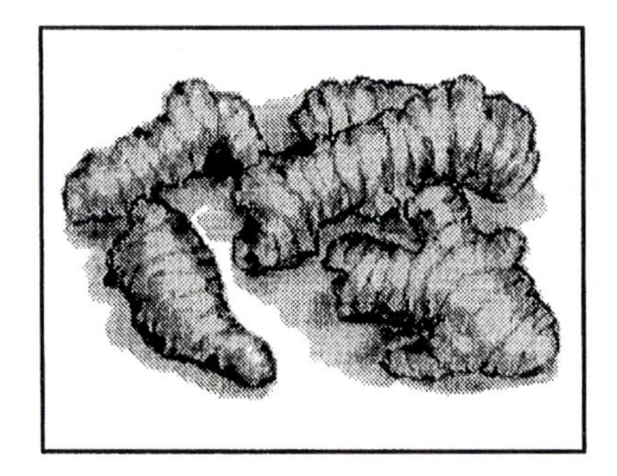

Ginger Root: Native to China, it was one of the first Oriental spices to be known in Europe. The ginger plant rhizomes send out many small "hands," so-called because of their irregular hand shape, which are dug up, washed, and sun dried, after which they are used as a spice or seasoning. The volatile oils, which give the odor, and a resin, which gives it pungency, have made ginger a favorite for use in baking and cooking. Ginger is used fresh, dried or crystallized with the fresh being the best flavored. Available year-round.

Glasswort: This coastal plant is also known as sea asparagus or marsh samphire. It is a green succulent plant that grows near salt marshes, resembles baby aloe, and is jointed together cactus-style. Glasswort is crunchy and salty like brined baby string beans. When young, it is crisp and pleasant tasting and often used in salads or as a garnish. It is also delicious boiled until soft, with the water drained away to reduce saltiness and then served with melted butter like asparagus. Not grown commercially.

GOURDS

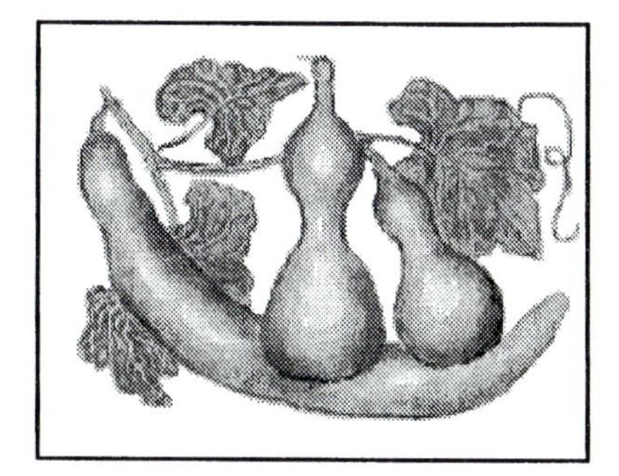

Gourd Bottle: This popular Chinese vegetable is also known as Calabash gourd, white-flowered gourd, Mao gua, or Poo gwa. Although originally native to Africa, there is some evidence of cultivation of this plant in South America in ancient times. It is possible that gourds may have floated there from Africa since experiments have shown that gourds will survive floating in seawater for more than 220 days with no loss of seed viability. Only the immature fruits are edible and they are used in soups and curries or stir-fried in the same manner as the hairy melons. The many varieties produce fruits of all sizes and shapes. Mature gourds and most round varieties are inedible due to their very bitter taste. They are principally grown for their hard, dried shells. Not grown commercially.

Gourd Snake: It is also known as serpent gourd or chicinda. This unusual cucumber-like gourd may grow to four feet or longer. They are green, streaked lengthwise with white, and when young, grow curling and twisting, turning bright red when ripe. A common practice is to tie a small rock to the base of the fruit to keep it straight. The fruit is only eaten when young. It is peeled, then sliced, boiled, and served as a vegetable. It has a very strong flavor but is more nutritious than many concurbits. Not grown commercially.

Gourd Wax: It is also known as ash pumpkin, Chinese fuzzy gourd, Chinese preserving melon, Tung kwa, Mo kwa dong gwa and Cham kwa. The wax gourd is a large, climbing gourd native to Java and now grown throughout eastern Asia. There is enormous variation in size. Usually found for sale in Chinese shops, the fruit ranges from the smallest, five to ten pounds and 12 inches in length, to the largest, three feet to five feet in length and weighing up to 110 pounds. The most important variety is called Cham kwa because of fruit shaped like a pillow. The gourd is green when mature and keeps in storage for up to a year. Wax gourds have a juicy texture and a mild flavor with white flesh. They are used as part of a sweet pickle relish stuffed with meat and vegetables and steamed, or cut into small chunks or slices for soups and stir-fries. Available year-round.

Gui Choy: Another member of the vast Brassica family, it is also known as Chinese mustard greens. Gui choy (Gai choy) is lime green, large, and twisted with a nappa-like bitter taste. It is used in salads, stir-fries, and as a separate vegetable. Available year-round from California.

H

HERBS

Herbs: Herbs are usually grown for their interesting associations, for their beauty and fragrance, and for their great importance to the art of cooking. To fully understand the relationship of herbs to food and a more precise definition of "herb," it is important to define that relationship to our everyday vegetables. The word "vegetable" came into existence about 200 years ago. Prior to that, all vegetables were known as "herbs" including all of our everyday vegetables such as carrots, beets, and cabbage.

What is the difference between an herb and a spice? Basically, herbs are the fragrant leaves of various annual and perennial plants that grow in temperate zones. They can generally be purchased fresh, dried in whole or ground form, or as a flavoring in vinegar or oil.

Spices are seasonings from the bark, buds, roots, seeds, and stems of various plants or trees. They can be purchased in whole or ground form. Today, herbs and spices, with their enticing piquant or aromatic qualities, serve mainly to enhance the flavor of food, whether in everyday cooking or gourmet dishes.

According to some cooking professionals, when cooking with herbs, the finer they are chopped, the more intense the flavor. Others say this isn't always true, arguing that even chopping can dilute their flavor. One tablespoon of fresh herbs equals one teaspoon of dried. For the most intense flavor, add herbs to soups and stews during the last 45 minutes of cooking, or save half and stir in during the final ten minutes before the dish is done. Some cooks would argue that 45 minutes is too long for fresh herbs; instead, add all fresh herbs 10 minutes before serving. When using herbs in thick, cold dishes such as dips and cheese spreads, refrigerate for several hours or overnight. To keep fresh herbs for any length of time, chop and then freeze them in ice cube trays filled with water. Then when you need seasoning, drop in a cube or two.

The produce industry today is responding to the increasingly sophisticated palates of consumers by offering an ever-increasing array of herbs. Listed here are just a few of the herbs that are appearing in the produce departments of your favorite food store and where they are used.

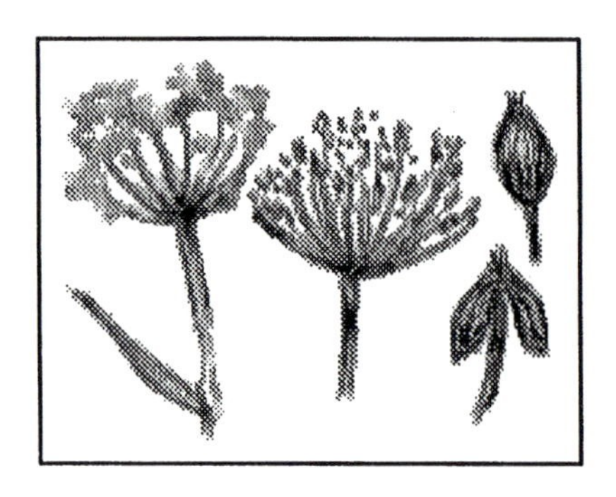

Herb Anise: This herb, also known as fennel, has a mild licorice flavor and is one of the traditional fish herbs. Leaves can be added to fish stews, soups, and casseroles. It also is excellent in lamb dishes, omelets, salads, and herb breads. Stalks can be used in any recipe that calls for celery or in creamy celery soup. Popular uses include seasoning pork, lamb, seafood, and beans. Available year-round.

Herb Arugula: Also known as roquette, this Mediterranean salad plant is a member of the mustard family. Arugula has a peppery taste that complements such Mediterranean foods as olives, garlic, tomatoes, peppers, and olive oil. Available year-round.

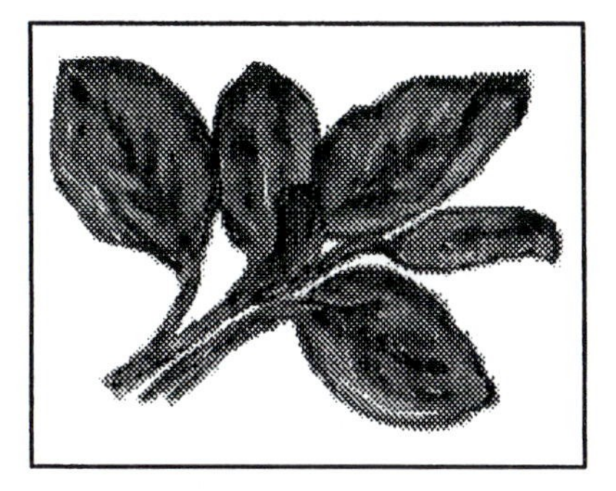

Herb Basil: This is a commonly used herb that adds a clove-like aroma and pungent taste to tomatoes, squash, cabbage, beans, pasta, pesto, poultry, or seafood. Leaves vary in color, from green to red-purple. Leaf size also is different, ranging from small, common basil leaves to the larger leaves of lettuce leaf basil. A young plant can be picked after six weeks. Do not store in refrigerator. Available year-round.

Herb Bay Leaves: With long, dull-green leaves about one half-inch wide, this aromatic and pungent herb is used to season soups, stews, and sauces. It is a traditional ingredient in split pea soup and spaghetti sauce. There are two types of bay leaves: Greek bay and California bay laurel. Available year-round.

Herb Chervil: This herb has a sweet taste, similar to tarragon. Add near the end of cooking to flavor stews, fish, and steamed vegetables. Chervil can be used in sauces calling for tarragon. Use in salads, salad dressings, meat dishes, savory sauces, egg dishes, or as a chopped garnish. Available year-round.

Herb Chives: A mild, onion-flavored herb, chives will enhance the flavor of almost any savory dish. Sprinkle liberally over fish, chicken, or egg dishes, or over a steaming baked potato. Chives also enhance almost any buttered vegetable such as carrots, beans, sweet corn, squash, peas, cauliflower, or mushrooms. Available year-round.

Herb Cilantro: Also known as coriander or Chinese or Mexican parsley, this herb has an assertive, sage-citrus flavor that can be compelling for some people. Use sparingly to season squash, eggplant, snow peas, and onions. Used extensively in Southwest fare such as guacamole, chili, salsa, and ceviche (cold seafood salad), cilantro also is important in Indian, Chinese, and Thai cuisine. Available year-round.

Herb Dill: The anise-parsley-celery flavor of dill goes well with fish, vegetables, soups, and salads. It is also delicious with poached salmon and in potato salad. Crown dill, a stronger-tasting item, is used for making pickles, while baby dill primarily is a seasoning. Dill plants have feathery leaves and clusters of small yellow flowers. When purchased fresh, plants should be selected on the basis of their resemblance to fresh salad greens. Dill seed also is a popular seasoning item. Available year-round.

Herb Marjoram: Sweet marjoram is a strong accenting herb used in egg dishes, soup, lamb, or vegetables. Its taste is similar to oregano, only milder. Like basil, marjoram is a member of the labiatae family. Plants are about 12 inches tall with oval-shaped, gray-green leaves and clusters of small white flowers. Available year-round.

Herb Mint: A sweet-flavored, aromatic herb, mint is a classic garnish and flavoring for summer drinks such as lemonade and punch, or a natural garnish for mint juleps, fruit platters, and frozen desserts. Mint also is a necessity in lamb dishes and many Middle Eastern dishes. Like all herbs, mint can be tossed in green salads or mixed into soft cheeses. It can be added to cooked carrots, green beans, peas, and beets. Available year-round.

Herb Oregano: Usually used to season Mexican, Italian, and Spanish dishes, oregano has a warm, aromatic scent and robust taste. Its uses include seasoning soups, stews, meat pies, pasta sauces, and shellfish. Available year-round.

Herb Parsley: Most commonly seen as a garnish, parsley adds a mild sweet flavor to foods. Often it is added to French, Italian, and Greek dishes. It can be chopped and added to soups, vegetables, meats, and sandwich fillings. There are two types of parsley: those with curly, fringed leaves, and Italian parsley, which has flat leaves. Available year-round.

Herb Rosemary: Spicy, strong, and fragrant, rosemary goes well in beef, pork, lamb, and veal dishes. Add to poultry, strongly flavored fish, and seafood. It enhances cauliflower, potatoes, eggplant, and peas. Available year-round.

Herb Sage: A common seasoning in meat, poultry, and cheese dishes, sage has gray-green leaves with a pebbly surface. Use sage sparingly, as the musty taste can be overpowering. Available year-round.

Herb Savory: Typically used in soups and as a meat and poultry seasoning, this herb has a slightly peppery flavor with a slightly sharp tang and can be used fresh or dried. Available year-round.

Herb Sorrel: With a sharp, lemony taste, sorrel resembles spinach but has pale green arrowhead-shaped leaves. Usually used in soups and sauces, sorrel also is used as a salad green or vegetable. The leaves can be cooked whole like spinach. Available year-round.

Herb Tarragon: An accenting herb used in mustard, tartar, and bernaisse sauces, and tarragon vinegar. A member of the sunflower family, tarragon has a sweet anise taste and should be used sparingly. One species, French tarragon, is primarily cultivated in the United States. Available year-round.

Herb Thyme: This herb is used as a spicy addition to Creole and Italian dishes and to season meat or poultry stuffings. When preparing game birds or roasts, thyme often is rubbed over the meat to season it. Available year-round.

Hinona-kabu: This elongated root is also known as the Japanese turnip. Although the turnip is native to Europe, this mild-tasting turnip is famous for its use in making the *Sakura zuke*, a Japanese cherry pickle so-called because of its red color. Its roots are approximately one to one and one-half inch in diameter. It can range from 10 to 12 inches in length. While the top one-third of the radish is red, the lower two-thirds are white. Hinona-kabu is used as a garnish, in salads, in cooking, and in making pickles. Availability limited in the United States.

Horseradish: Native to eastern Europe and western Asia, this plant is widely grown for its pungent root. The root is generally 6 to 12 inches in length and 1 to 3 inches in diameter, with several protuberances at one end. The skin is the color and texture of a dirty, wrinkled, gnarled parsnip or parsley root and should be very hard and free of soft spots. The root is usually grated, and combined with oil, vinegar, or cream to make a sauce that is used as a seasoning for vegetables, relish, meat and kielbasa, and seafood salads. In the Middle Ages, horseradish was used to mask the flavor of spoiled meat. The flavor is pungent, very hot, and should be used sparingly. Available year-round.

Houttuynia: Houttuynia is native to eastern Asia and China. The young shoots and heart-shaped leaves are picked early in the spring and either eaten raw in salads or cooked like spinach. A wild form (yerba mansa), formerly used medicinally for diseases of the skin and blood, grows in eastern California, Texas, and Mexico. Not grown commercially.

J

Jerusalem Artichoke: Also known as Sunchoke® and girasole, it is one of the few native plants of northern North America to take hold in the Old World. Introduced to France in the seventeenth century, it was an immediate success. No one is really sure where it received the misnomer Jerusalem artichoke since it is not a member of the Compositae (thistle) family, nor does it have any connection with Jerusalem. Jerusalem artichokes are now being marketed under the names of Sunchoke® in an effort to rekindle interest in this once popular tuber. The scientific name is *Helianthus tuberosus*, meaning sunflower. This tuber of a sunflower, tan to cream in skin color, smooth and pearly (can be lumpy and bulbous) in shape, has an ivory flesh that is extremely crisp, like a water chestnut and a flavor that is sweetly fresh. It can be grated, sliced, diced, or julienned for use in salads, meat dishes, or fish dishes. A one-half cup serving has about 60 calories and contains iron, phosphorous, and B vitamins. Available year-round.

Jicama: Also known as Mexican potato, yam bean, Mexican yam bean, Ahipa, and Saa got, it is native to Mexico and to the headwater area of the Amazon River in South America. Introduced to the Philippines from its source by the Spanish in the seventeenth century, it became a favorite crop of Chinese gardeners and spread throughout Asia and the Pacific. The jicama is brown skinned, one-half to six pounds in size, with the shape of a large turnip. The crisp, white flesh has a delicate, sweet flavor. Jicama can be served raw, sliced, or served as part of a relish tray, or in stews, soups, salads, and as a substitute for water chestnuts. It is very low in sodium, a fair source of vitamin C, and a three and one-half ounce serving contains about 45 calories. Available year-round.

The immature jicama pods can be eaten, but the mature vines, beans, and pods are poisonous, containing the insecticide and fish poison rotenone.

K

Kale: Also known as curly kale, kail, and borecole, it is native to Europe and the Mediterranean area. This crinkly leaved, non-heading cabbage was one of the earliest members of the cabbage family to be cultivated. It was introduced to North America by early European settlers. A favorite of the Scots, Germans and Scandinavians were also especially fond of lake. In America, it quickly became a favorite vegetable in the South. Kale leaves are bluish-green with a mild cabbage flavor. It is as versatile as cabbage or spinach, and can be used in salads or soups and as a separate vegetable. Kale is an excellent source of vitamins A and C (twice the RDA). A three and one-half ounce serving contains about 53 calories. Available year-round.

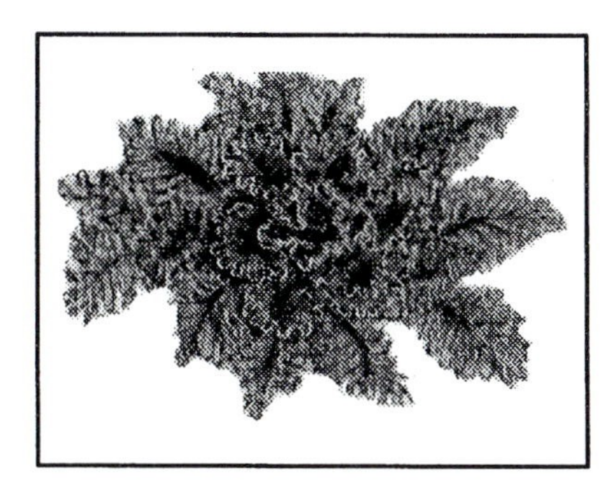

Kale Flowering: This beautiful plant is also known as flowering cole, flowering cabbage, ornamental kale, and salad savoy. It is the same species as regular kale except (evidently through select breeding) in appearance. Flowering kale is a fantastic display of loose, ruffle-edged leaves of cream, violet, or maroon, veined and marked with designs of one or all of the other colors that branch out from a central stock. Its subtle hues and fluffy forms make it an ideal item for filling out table displays, formal cornucopias, for holding containers of dips, and as a centerpiece. Available year-round.

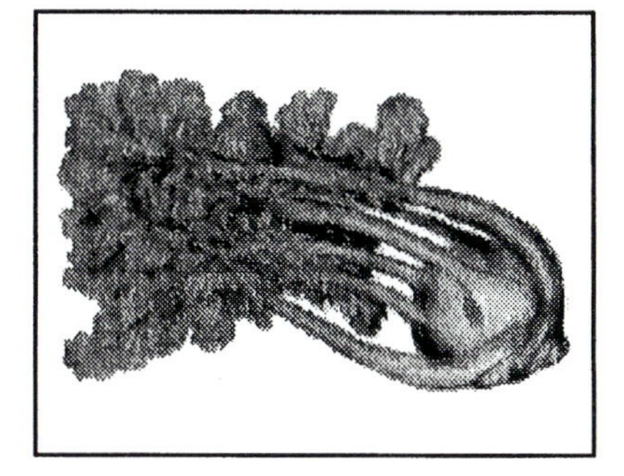

Kohlrabi: Another unusual vegetable from the Brassica family, kohlrabi is also known as cabbage turnip and is sometimes called the "educated turnip." Kohlrabi is not a cross between cabbage and turnip as some think, but a separate vegetable that belongs to the same family. There are two varieties: green and purple. Characterized by a swollen stem and a delicate, kale, broccoli-like flavor, kohlrabi is a green vegetable that is totally edible. Kohlrabi leaves are used in salads, cooked, steamed, and served as a vegetable dish. The stems (bulbs) can be used raw or cooked like turnips. Kohlrabi is high in potassium and vitamin C, and contains about 40 calories per cup. Available year-round.

Komatsuna: Also known as mustard spinach or spinach mustard, komatsuna is similar in appearance and taste to regular mustard greens although they are actually forms of turnips developed for their leaves and leaf-stalks. Komatsunas may be eaten at any stage, and the flavour has been described as midway between cabbage and mustard. The young shoots can be used in salads. Leaves can be boiled lightly or steamed, stir-fried, or combined with other greens and used in soups. Available year-round.

Korila: Also known as schrad, caygua, and achocca, it is a climbing gourd native to Central and South America. The fruits are oval, pale green, flattened, and hollow for much of their length. They can be smooth or covered in soft spines. They open explosively, ejecting black seeds. The taste is similar to a cucumber, and the fruit is eaten raw or cooked. Not grown commercially.

Kudzu: Native to China and Japan, this plant is an uncontrolled climber. Long cultivated in the Philippines and Polynesia, it has been introduced to the southeastern United States, where it is now threatening to become a pest. The leaves can be used as a salad green and the tubers (weighing up to 15 pounds) can be used in cooking. The tubers are starchy and more often used as a source of starch for thickening soup. Available year-round.

L

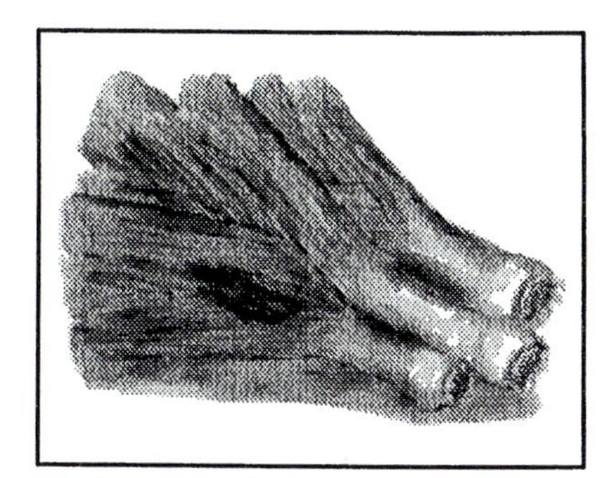

Leek: Like kohlrabi, the leek's origin has been lost in time. The cultivated variety is a mild-flavored member of the onion and lily (Alliaceae) family. It is an outstanding vegetable for use in omelets, soups, salads, or combined with other vegetables. It also is excellent served as a separate vegetable alone, or with a cream sauce. Three of the more popular commercial varieties are the American Flag, Blue Leaf, and Carentan.

There is a wild variety (called "ramp" in some areas) that is native to southeast Canada through New England to Georgia. There is even a ramp festival in south-central West Virginia. Having a stalk streaked with violet and leaves that resemble those of lily of the valley, it has a very strong onion-garlic flavor. Use in the same manner as the cultivated leeks, except discretion is advised because of its stronger flavor. Leeks are a good source of potassium, vitamin A, and vitamin C, and a three and one-half ounce serving of raw leeks contains 52 calories. Cultivated leeks are available year-round, while wild leeks are available March to June.

In legend, before the battle in which the Welsh won a victory over the Saxons in 640 AD, Saint David advised the Welsh to wear leeks from a nearby garden in their caps so they could identify their enemy and not kill one of their own soldiers.

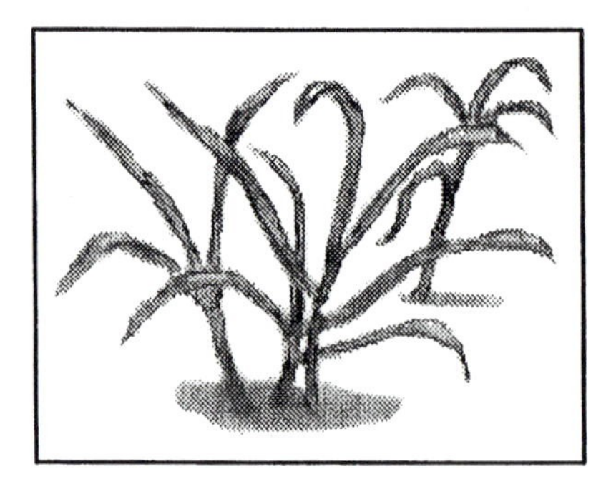

Lemongrass: It is also known as Sereh, Takkrai, citronella root, and Ti de lemon and is native to Southeast Asia. A tropical perennial, lemongrass has long, woody stalks with white root ends. The leaves and stalks are green, similar to green onions, and the texture is dry, brittle, with a lemony flavor. It is inedible unless peeled. Peel about an inch from root end and pull back three to four layers until the tender white portion is exposed. Used for seasoning in salad dressings, chicken and fish dishes, marinades, beverages, and condiments. Available year-round.

LETTUCE

Lettuce: Native to Europe and the Mediterranean area, lettuce was being served to the Persian Kings as early as 55 BC. Romans, during the time of Augustus, ate lettuce in salads at the end of their meals. Most traditional Italian meals still serve salad after the main meal. Lettuce migrated to China near the end of the ninth century where today, the loose-leaf variety still remains the most popular along with the stem variety called celtuce. Various kinds of green, red, and spotted lettuce in loose forms have been used for centuries; the heading types having been developed only in recent times. The name comes from Latin words referring to its milky juice. Ranging from upright, loose-heading types, loose broad-leaf types, semi-heading types, to firm, tight-headed types, lettuce has ruled supreme as the base for almost all salads. In addition to its almost unlimited use in salads and appetizer dishes, lettuce makes an excellent soup when combined with bouillon. Available year-round.

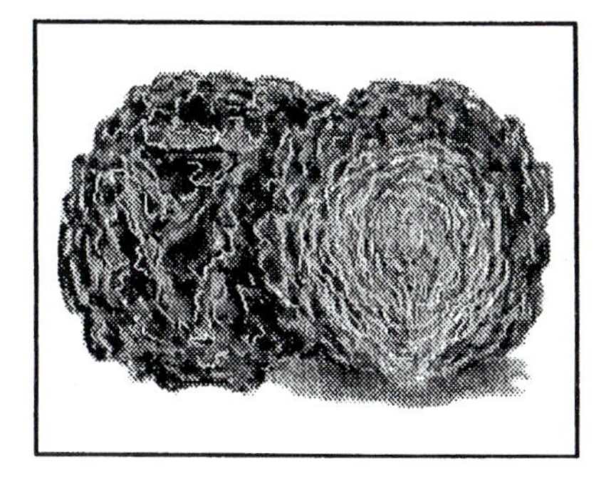

Lettuce Iceberg: It is also known as crisping lettuce or head lettuce. The beginning of the head lettuce industry began in 1896 with the introduction of crisphead lettuce. This strain, particularly suited to the climate of the Southwestern United States, quickly became the base for large-scale commercial production of lettuce. From less than 600 acres in production in Los Angeles County in 1910, the acreage rapidly expanded to almost 19,000 acres within ten years. As Los Angeles County increased in population and available farm acreage decreased, the migration of the lettuce business moved north until it reached the Salinas Valley. The perfect weather and available land in the valley enabled California to become the leading producer of lettuce in the United States.

According to California historian Burton Anderson, the success of head lettuce in California was due to three factors: ideal climate for growing, the invention of the refrigerated rail car so lettuce could be transported long distances to the markets, and the shippers aggressively marketing their product under the name "Ice Berg" or "Western Ice Berg" lettuce.

This solid head, tight-leafed, mild-flavored, crisp-textured lettuce is by far the most popular. In addition to being a source of fiber, lettuce is a good source of vitamin A and is low in calories. It is excellent for use as a salad base, for appetizer dishes, and for adding texture to all types of sandwiches. Iceberg lettuce is available year-round with California, Arizona, and Florida being the major producers.

Lettuce Butterhead: This soft leaf lettuce is also known as Boston, bibb, and limestone. These semi-heading lettuces are named for their delicate almost buttery flavor. They are small-headed types with soft, light-green leaves, and thinner, less-prominent veins in their leaves.

Butterhead lettuce is a welcome addition to salads and sandwiches. Available year-round.

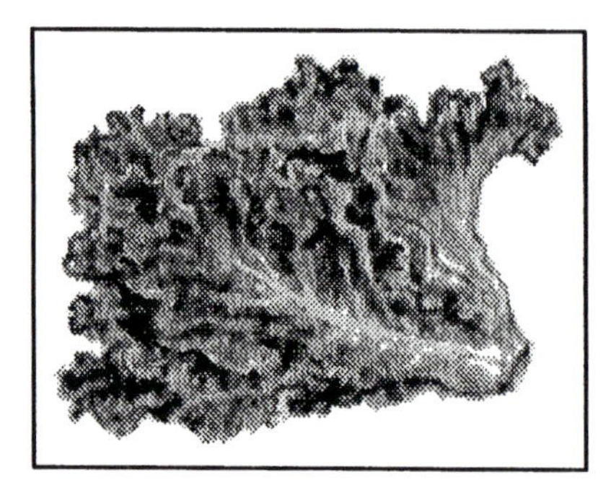

Lettuce Loose-leaf: Also known as red leaf or green leaf, these loose-heading, broad-leafed lettuces take their names from the way they grow and the way the leaves branch out from the stems rather than forming heads. Colors range from red, bronze, dark and light green, to chartreuse, and textures from smooth and puckered to ruffly and frilled. Used in salads and sandwiches as well as garnish. Available year-round.

Lettuce Romaine: It is also known as Cos lettuce. Coming from the Greek island of Cos, romaine is a loose-leaf, loose-heading lettuce, with big, crunchy leaves characterized by their upright stance. Romaine is an excellent addition to almost any vegetable salad. Available year-round, with California and Arizona the two major producers.

Lo Bok: Also called Mooli, Luo bo, and Daikon, it is one of the many varieties of Oriental radishes. The lo bok is shorter and squatter than daikon, five to ten inches in length, and approximately two inches in diameter with a sharp, hot radish flavor. The flesh and skin are white with some greening near the leaf end. A staple item throughout the Orient, it is used with broiled fish, salads, pickling, and seasonings. Available year-round from California.

Lotus Root: This beautiful, well-known waterplant is also known as lily root, water lily, sacred lotus, Indian lotus, water lotus, Leen ngua, and Hasu no. Lotus root is an ancient vegetable, a member of the Water Lily family, and is cultivated as a pond plant from rhizomes. It has been traced back as far as 708 BC and is now cultivated in seven countries including the United States. Lotus roots are ivory to brown-colored smooth stalks that look like a string of fat sausages, six to ten inches long, two to three inches in diameter, separated by narrow necks. The interior has a pattern of holes when cut widthwise that looks like a snowflake. Lotus roots are mildly sweet, slightly astringent, with a nutty artichoke overtone. They are baked, boiled, fried like a potato, and added to soups, stews, and vegetable dishes. In India, it is broiled, mashed, and used in chutneys. Available year-round from China.

There is a second species, Nelumbo lutea, that is native to North America. It is found growing in ponds and estuaries from Florida and Texas to as far north as New York, southern Ontario, and Minnesota. Its seeds and tubers are edible.

M

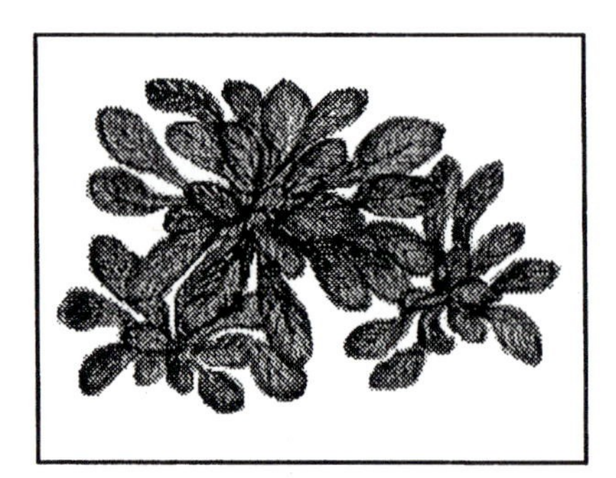

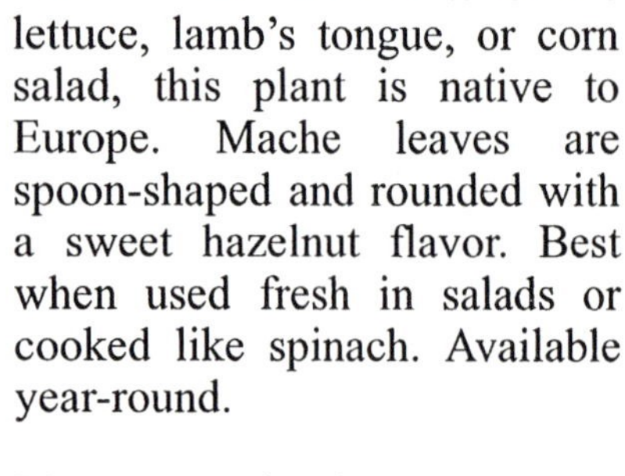

Mache: Also known as lamb's lettuce, lamb's tongue, or corn salad, this plant is native to Europe. Mache leaves are spoon-shaped and rounded with a sweet hazelnut flavor. Best when used fresh in salads or cooked like spinach. Available year-round.

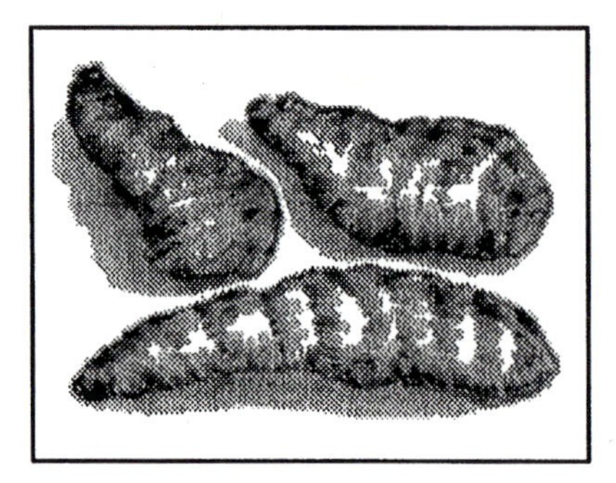

Malanga: Also known as yautia, tannia, malanga amarilla, or cocoyam, it is native to the American tropics. The 40 or so species include some of the oldest root crops in the world. Sometimes confused with taro root, malanga may be shaped like a long yam or curved like a short club. Ranging from one-half pound to two or more pounds, it is shaggy, brown, with patchy, thin skin and beige, yellowish, or reddish flesh that has an extremely crisp, slippery texture. The flavor has been described as slightly earthy and nut-like. Malanga is used as a potato substitute and a soup and stew thickener. The tubers are peeled and the product is cooked like potatoes. Malangas are a good source of thiamin, riboflavin, vitamin C, and iron, and a one-half cup serving of cooked malanga contains about 135 calories. Available year-round from Florida.

Mallow: Mallow plants formerly grown in Europe are still common in China today. An easily grown annual, mallows form rosettes of long-stemmed leaves before sending up a tall flowering stem. All parts of the plant are edible when young, and have a mucilaginous texture with a rich, pleasant taste. The young leaves and shoots are picked as required and eaten in soups and salads or lightly boiled like spinach. Available from California.

Mibuna Greens: It is also known as mibu greens. A recent import from Japan, it appears to be related to mizuna. It has a long history of cultivation in the Japanese prefecture of Kyoto. It is a mild flavored vegetable characterized by long, narrow leaves, with some hybrids producing plants up to six pounds in weight. Mibuna is used in stir-fries, steamed or boiled as a vegetable, and used fresh in salads or in pickling. Availability is limited due to shortages created by lack of production.

Mitsuba: This small plant is also known as Japanese wild chervil, San ye qin, and San ip. Growing wild in Japan, mitsuba is clearly a Japanese plant. Considered as one of its most important fresh vegetables, mitsuba is found in many traditional Japanese dishes. Mitsuba, growing about 12 inches tall, with dark green, heart-shaped leaves with saw-tooth edges, has a mild flavor, often described as celery-like with an overtone of parsley. Leaves and stems are used in soups, rice dishes, custards, or as a garnish. They also make attractive, dark green seedling sprouts with a light, pleasant flavor. Available year-round.

Mizuna: It is also known as Kyona, Mizuna mustard, or pot-herb mustard. Cultivated since ancient times in Japan, mizuna is a small green that is closely related to the leafy turnip and commonly grown in Japan. They are fast growing, forming clumps of finely dissected leaves, which stand for long periods of time without getting tough. Mizuna is used in salads, soups, steamed, stir-fried alone or with other vegetables, or with meat, poultry, or fish. Available from California.

MUSHROOMS

Mushroom: The mushroom is not a vegetable but a fungus that has no chlorophyll for manufacturing its own food. Consequently, it lives on decaying organic matter. Almost all commercial mushrooms sold in the United States and Canada are of the same cultivated variety, *Agaricus campestris*. Some strains are cream colored, others brown, but most are white.

Mushrooms have been considered a delicacy since the time of the pharaohs. They appear in Egyptian hieroglyphics as food for the pharaohs who considered them too delicate for common people to eat. They were favored in ancient Rome as a "food of the gods." Cultivated by the French in the seventeenth century, they were introduced to North America as a cultivated plant around 1890. By 1920, mushroom farming was underway in the United States.

Mushrooms are very versatile. They are used for soups, sauces, garnishes, gravies, stuffings, and raw in salads, and they combine well with peas, beans, and other vegetables. To keep mushrooms white, dip them in boiling water a few seconds before cooking.

Mushrooms are rich in amino acids and a good source of protein, iron, copper, riboflavin, and nicotinic acid, and a fair source of vitamins B1, C, and K.

Although there are more than 2,000 species of mushrooms eaten throughout the world, we will only relate to those most commonly sold in the North American markets. Some of the more popular varieties are listed here.

Of the many wild mushroom varieties available, a few are poisonous, so it is wise to leave gathering of wild mushrooms to the experts.

Mushroom Agaricus bisporus: It is also known as the button mushroom. Although there are three different strains, their appearance and qualities are similar, differing only in color, with white, off-white, and dark brown being the most common.

Mushroom Black Chanterelle: A wild mushroom from the forests of the Pacific Northwest, it has an earthy flavor, is grayish black in color, and fluted in shape. It is best when eaten cooked. Available January through March from Oregon.

Mushroom Chanterelle: These mushrooms grow wild in the forests of the Pacific Northwest, but in the past few years they have been cultivated to some extent. Most common is the yellow chanterelle, which should be cooked before eaten. It resembles a curved trumpet with ribs, has a fruit-like aroma with a delicate flavor, and is highly favored as a cooking mushroom. Available September through April from California, Oregon, and Washington.

Mushroom Chicken of the Wood: A large mushroom with a fan shape. The flesh is white or salmon orange, with deep orange to yellow skin. It must be cooked. The young mushrooms have the best flavor. Available late summer and fall from California.

Mushroom Crimini: It is also known as Italian or California brown and it is a revival of the California brown mushroom. A large mushroom and cousin to the white *Agaricus*, the crimini is light tan to dark brown in color and has an earthier, spicier flavor than the white button mushroom. They are excellent for stuffing.

Mushroom Enoki: Some experts state it is native to Japan and others state it is native to North American forests. Leaving the question to be answered by botanists, the enoki is only the cultivated form (which looks different from the wild) that is presently marketed. It is now being produced commercially in California. Unlike other mushrooms, the enoki is harvested and packaged in clumps. The enoki is a creamy-white mushroom with long, slender stems topped with small round caps. It is a mild, sharp-flavored mushroom that adds an elegant touch to salads, soups, stir-fries, and other dishes. Available year-round from California.

Mushroom Golden Mountain Oyster: A fan-shaped mushroom, light brown in color, with a meaty flavor. It is available year-round from Utah.

Mushroom Hedgehog: This mushroom has a buff-orange cap with white tooth-like projections under it and it has a sweet, hearty mushroom flavor. Available December through March from Oregon and California.

Mushroom Hon Shemeji: This mushroom is relatively new to the United States. Grown commercially in Japan, it comes in clusters with small caps and has a lobster-like flavor. The entire mushroom is edible: the cap, stem, and most of the cluster base. Available from California.

Mushroom Lobster: This mushroom has a slight fish taste, ranges from orange to red in color, and looks like paper mache. It is good for soups and stews. Available mid-July through August from Oregon.

Mushroom Morel: A wild mushroom primarily from the Pacific Northwest, the morel is also grown in Wisconsin and Michigan. It has a shape resembling a pine cone, a cap with a honeycomb texture, and a light tan to brown to black color. It has a strong, earthy flavor and cannot be eaten raw. Morels can be used in soups, stews, and pastas, as well as meat, fish, and poultry dishes. Available late April through June.

Mushroom Pleurotus: Also known as oyster mushroom, the pleurotus is one of the more exotic mushrooms offered on the market today. A gray-brown mushroom with a delicate flavor and texture, it is used with meat, omelets, and stews. It can be used raw, but tastes better when cooked. Available year-round from California.

Mushroom Porcini: It is also known as cepes or boletus. This mushroom has a thick stem and a smooth, round, brown cap lined with a closed-texture network of shallow tubes, which are edible when the mushroom is in good condition. It has a chewy, meaty texture with a beefy, nutty flavor, and should be used as fresh as possible.

Mushroom Portobello: This mushroom is nothing more than the common brown *Agaricus bisporus*, formerly known as the Italian brown or California brown. These small button mushrooms were renamed crimini not long ago but now they have found a new life under the sexier name of portobello. Although some people insist on their spelling as "portabella" or "portabello," the spelling "portobello" (meaning beautiful port in Italian) makes the best linguistic sense. These mushrooms are of substantial size due to their longer growing period. Given an extra five or six days of growth, a crimini mushroom expands dramatically, develops flavor, and opens its inky gills–a portobello is born. Due to this longer growing period, portobello mushrooms take on a darker, richer flavor, which is unique to them.

Stretching four to six inches across the cap, with a small stem end, they have a brown, woodsy hue and an earthy, unmistakable mushroom aroma. Basically unknown four or five years ago, they have now become one of the most popular mushrooms in the produce industry.

The ability of portobellos to be a meat substitute probably accounts for much of their popularity. Expert chefs say they have the texture, the body, and even the look of a nice steak. And steak-like, they take to the grill or broiler, losing only about 20 percent in water weight during cooking. Portobellos can be sautéed, served as an appetizer, used as a meat substitute, and even used as a pizza shell. Just turn it upside down, pop out the stem, fill it with tomato sauce, cheese, or whatever you like on pizza and bake it until bubbly. They are low in calories and contain no fat, no cholesterol, and no sodium. They have a trace of vitamin C and iron. Portobellos are available year-round.

Mushroom Shitake: In Asian countries, shitake is called the "elixir of life." These mushrooms were only found growing on trunks of oak trees. Today they are cultivated in temperature-controlled environments. A large, umbrella-shaped mushroom, shitakes are brown-black in color. The meaty flavor makes them a good choice for salads, meat dishes and sauces, whether raw or cooked. Available year-round from California.

Mushroom Wood Ear: It is also known as Judas Ear and Mook yee. A mushroom that looks like flattened plates with occasional whorls and ruching (similar in form and veining to an ear), wood ears can be the size of a quarter or a teacup saucer. They have an earthy flavor and a springy, soft texture. Available year-round.

Mustard Green: It is also known as curled mustard, mustard spinach, Indian mustard or leaf mustard. Botanists still debate whether this member of the cabbage family originated in Siberia, Europe, or India. Sharp and pungent, mustard greens are the most powerful of the bitter greens. Several commercial varieties are available with leaf colors varying from dark green to light green, all with a prominent stem, crinkly edges, and peppery flavor. Mustard greens are used raw in salads, in Oriental cooking, and as a separate cooked vegetable. They are also a favored food in the southern United States. Mustard greens are low in calories, about 20 calories per cup when cooked, and they are an excellent source of vitamins A and C, calcium, and potassium. Available year-round from California.

Mustard Chinese: It is also known as brown mustard or Kai tsoi. Several varieties are now available, including some Japanese varieties such as Osaka Purple and Red Chirimen (giant-leaved) and Big Heart and Chicken Heart (large-headed).

Mustards are not to everyone's taste. They are identified by a peppery flavour and the degree of pungency varies enormously from one plant to another, even from one part of a plant to another part. Chinese mustard has thickened stems like celtuce and simple, oval leaves. The plant is generally eaten cooked, steamed, or fresh in salads. Mustards have high levels of vitamins A and C and are a good source of calcium, iron, potash, and phosphorus. Available year-round.

N

Nappa: It is also known as Chinese cabbage, Pe-tsai, Wong bok, Hakusai, Siu choy, or Chinese leaf. There are three main types of Chinese cabbage: the barrel (hearted), the tall cylindrical, and the loose-leaf type. Nappa is a barrel type and a celery cabbage, with a crisp, crunchy texture, mild radishy flavor, and long, white, wide, pale ribs that fan out into crinkled, veined leaves tinged with light green. It is a favorite vegetable for use in all manner of Oriental cooking. In China, Japan, and Korea, Chinese cabbage is a standard part of meals in the form of salted or spiced pickles. Nappa is very low in calories, about 16 per cup cooked, and is a good source of folic acid, potassium, vitamin A, vitamin C, and is very low in sodium. Available year-round.

O

Oca: Also known as iribia, cuiba, and New Zealand yams, it is a common crop in the Andes and grown in the same fields as potatoes. Its cultivation ranges from Venezuela to Argentina and it is a staple in parts of Bolivia. It is also popular now in parts of New Zealand after being introduced from Chile in 1869. Tubers grow from three to six inches in length, with elongated to round appearance and colors varying from white, yellow, and piebald to black. The flavor is similar to potatoes with slightly sour overtones caused by oxalic acid. The acid can be reduced by leaving the tubers in the sun for a few days. They then become floury and sweet. Available from March through September.

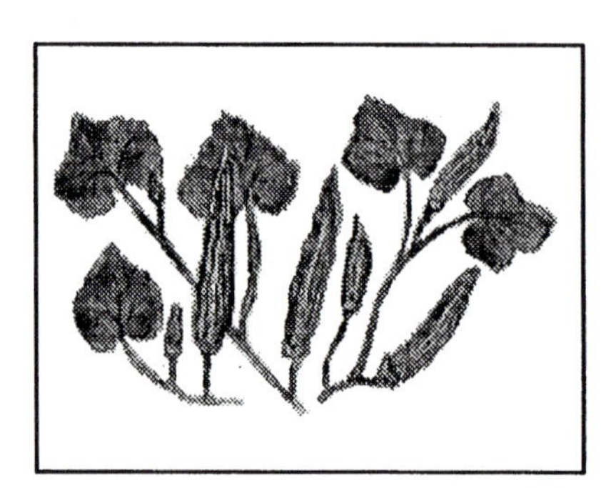

Okra: Thought to have originated in northern Africa somewhere near Ethiopia, it came to North America's South through the slave trade, and it is still a staple in Creole cooking. Its name comes from the Gold Coast of Africa language Twi, as *nkruman,* gradually abridged to "okra." The name *ngumbo* from slaves of Angola gradually became "gumbo," which is still widely used as a synonym for okra and the many dishes derived from okra. The elongated pods (two to seven inches long), light green in color, and tender, have a sweet taste and when cooked; the texture is oftentimes slippery and mucilaginous in some dishes, though it can also be prepared to be nutty and crunchy. To keep okra from turning black while cooking, cook okra in a porcelain, glass, or enameled vessel and not in metal vessels like an iron skillet or copper pan. Okra can be steamed, boiled, pickled, sautéed, deep-fried, or braised or added to salads or soup. Okra is a good source of vitamins A and C and provides a modest amount of folic acid and calcium. Available year-round.

ONIONS

Onions: Although native in wild form to Europe, Asia, and North America, this is the vegetable that gave Chicago its name. For Chicago is the area where Chippewa Indians harvested these *she-gua-ga-winshe* growing wild.

For centuries, onions have been credited with everything from making hair grow on bald heads to giving valor to the Greek warriors. The name "onion" comes from the Latin *unio*, from the French *oignon* and the English *unyun*. Onions belong to the same family as the narcissus. While dry onions are low in calories and practically everything else, they do provide essential minerals, and green onions are especially high in vitamin A. Their uses include salads, soups, stews, sandwiches, as a separate vegetable, and for seasoning all kinds of other foods.

Onions usually fall into two categories: spring/early summer fresh and late summer/fall storage. Spring/early summer fresh onions are Granex, Grano, and hybrids. These are yellow, white, or red, with varied shapes from flat, top-shaped, and round. These mild, early-maturing varieties are planted in late fall, winter, or early spring for spring/summer harvest. This type of onion is often used raw in salads, and some are mild enough to eat-out-of hand.

The late summer/fall storage onion includes a range of variety, mostly globe and Sweet Spanish hybrids. These are available in white, yellow, and red color, round-shaped, and come in all sizes. The strong, full-flavored globe varieties are noted for their excellent storing and cooking qualities. Sweet Spanish hybrids often grow to three to five inches in diameter and are plump in appearance. Sweet onion varieties such as Maui Sweets, Walla Walla Sweets, Texas 1015 Supersweet, and the Vidalia have a milder flavor, higher sugar content, and are served fresh more often than the stronger-flavored onions.

Most varieties of globe onions are round and brown-skinned when cured, as well as the leading variety sold in most food stores. Spanish-type onions are the largest, usually brown-skinned, and the preferred type for slicing due to their sweet flavor. Red slicing types are large, flat or semi-flat, and popular as a color addition in salads and as a garnish. Grano/Granex/Bermuda onions are shaped usually like a toy top, fairly large in size, good for slicing, and represent a good portion of the onion production in the yellow, red, and white varieties. Boiling types are small, usually white or yellow, round, and used in stews, vegetable dishes, and with pot roasts.

Long before the Christian era, the onion was worshipped in Egypt. Onions as well as garlic were believed to have power to overcome evil and to prevent diseases. Immense amounts of gold were spent to provide onions and garlic to the builders of the great pyramids.

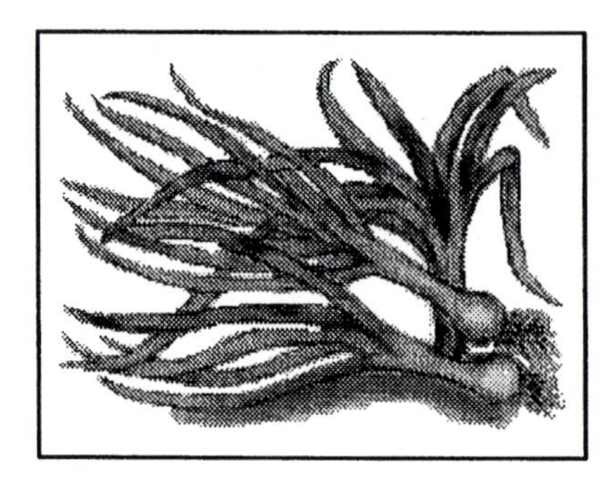

Onion Green: Green onions are also known as bunching onions. These are white onion varieties that are harvested when they are very immature for use in salads and cooking. They come in two types: those with and those without a small bulb. Available year-round.

Onion Imperial: These yellow onions are grown in Southern California's Imperial Valley. Predominantly single-centered with thick rings, the Sweet Imperial has dry, tissue-thin skin and a short tight neck. A mild, sweet tasting onion that is used in salads, sandwiches, garnishes and soups, it is available April through mid-June from California.

Onion Maui Sweet®: These are mild, sweet onions grown in the Kula agricultural district on the island of Maui. The rich, fertile farming area ranges in elevation from 2,000 to 5,000 feet above sea level. Kula is located midway up the slope of Haleakala, an ancient volcano on Maui. The sweetness of the onion is a result of the warm days, cool evenings, and volcanic soil. The farmers use both Granex and 1015 seed, depending on the season and weather conditions. Maui Kula onions can be eaten raw, used in salads, or used in regular cooking. Available year-round from Hawaii.

Onion Pearl: These tiny onions (about the size of a large pearl, hence the name) have a mild flavor, crisp texture, and come in three colors: white, red, and gold. They are used whole or sliced in salads, vegetable dishes, casseroles, sauces, and meat dishes. They can be skewered for shish kabobs, and they are often pickled or marinated. Available year-round from California. A wild onion known as apaz, looks and tastes similar to the pearl onion and is available from Oregon in June.

Onion Texas 1015 Supersweet: Named after its recommended planting date of October 15, this yellow Grano variety is an extremely mild onion that tends to grow to jumbo sizes. It is predominately single-centered. It holds up well in storage (up to six months with proper care) and is available April through June.

Onion Vidalia®: A Georgia-grown, yellow Granex hybrid known for its sweet, mild flavor, the Vidalia was first grown in Toombs County, Georgia in 1931. It has a golden-brown bulb with a white interior and does not store well, lasting usually one to two weeks. Vidalia onions are excellent in salads, soups, sandwiches, and meat dishes. Available April through June from Georgia.

Onion Walla Walla Sweet: Named after the area where it is grown, the Walla Walla Sweet onion is thought to have originated on the island of Corsica. Discovered by the French soldier Peter Pieri nearly a century ago, it was brought by him to Walla Walla, Washington in the late 1800s. The onion was developed over several generations through careful selections of onions from each year's crop. This has a golden-brown bulb and white interior and is well known in the Pacific Northwest for its mild sweet flavor. Like the Vidalia onion, Walla Walla Sweets do not store well. They are excellent in salads, soups, stews, sandwiches, and meat dishes. They are a good source of vitamin C, are sodium free, contain no cholesterol, and one medium-size onion only contains 60 calories. Available June through August from Washington State.

Onion Welsh: It is also known as Japanese bunching onion, Chinese small onion, or scallion. The Welsh onion is similar to large, very coarse chives, with hollow green leaves and almost no bulb at the base. It is widely used in Chinese and Japanese cooking as a flavoring in fried vegetables. It may also be used as a winter substitute for chives. Available from California.

Orache: Orache is native to central Asia and Siberia. It is grown as a substitute for spinach. It is hardier and more easily grown than spinach and has a stronger flavor. The green, red, and yellow varieties are still cultivated, and the lower leaves are large, heart-shaped, or triangular. Its uses are the same as spinach. Available from California.

P

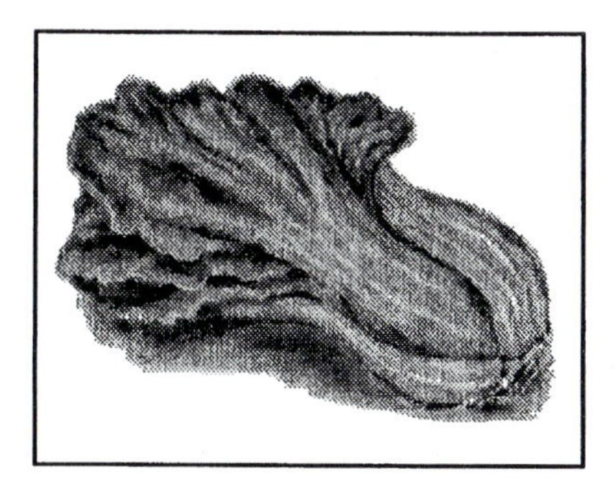

Pak Choi: It is also known as Quing cai, Chinese celery cabbage, Chinese white cabbage, and mustard cabbage. Older than nappa, this plant has been cultivated in China since the fifth century AD. There are several species of the Pak choi family. One is the unique rosette Pak choi. A small plant with a rosette of upright leaves, thickened and flattened with white petioles, and smooth, rounded blades. Other varieties include the Chinese white pak type, soup spoon type, green leaf type, and the squat or canton type. Pak choi is best used fresh since it does not keep well under normal conditions. Pak choi is usually chopped before cooking, but small plants may be cooked whole. Used in stir-fries, it is also added to soups, to meat, chicken, fish, and noodle dishes, and to salads. It is picked and sometimes served as a separate vegetable. Available year-round.

Parsley: Native to north and central Europe, parsley is more common in home gardens around the world than any other herb. A short-lived perennial with more or less curly and feathered leaves, it is unequaled as garnish. Fresh, chopped, dried, or powdered, the leaves have endless uses as seasoning in sauces, dressings, croquettes, and in egg, fish, shellfish, and meat dishes. Available year-round.

Ancient Greeks crowned their brows with parsley leaves in the belief the leaves would stimulate and awaken the brain.

Parsley Root: Also known as rooted parsley, Dutch parsley, and petoushka, it is native to northern Europe. Parsley root is an unusual member of the Umbelliferous family, it is cultivated for its roots, rather than its leaves. The small, irregular-shaped root (often double-rooted), looks like a small parsnip attached to large feathery leaves. The taste, somewhere between celeriac and carrots, adds flavor and aroma to braises, stews, and soups. Available from New Jersey during the summer months and from California year-round.

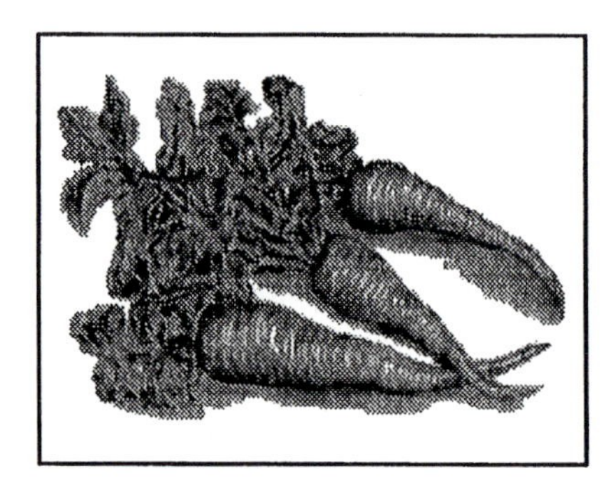

Parsnip: Native to the eastern Mediterranean area, parsnips were probably known to the early Greeks and Romans in their wild form. This vegetable is a taproot of the Umbelliferae family and cousin to the carrot. The typical parsnip is white to cream in color, 6 to 12 inches in length, and shaped like a tapered carrot having a sweet nut-like flavor, they are excellent boiled, fried, in stews, in fritters, French fried, glazed, and creamed. A good wine can even be made from them. They are rich in potassium, low in sodium, and have a moderate energy value. Available year-round from California and the Eastern states.

PEAS

Pea Green: Also known as the English pea, it is native to the Middle East. Peas have been around for thousands of years. In fact, peas found at archeological sites in Burma and Thailand have been carbon-dated to 9750 BC. First grown only for the dry seeds to be used in soups, fresh green peas were considered a delicacy until the seventeenth century when they came into vogue among the members of the French court. Many new strains have been developed over the years, especially in England, where they thrive in the cool, moist climate. The English pea has a large pod that bulges away from the pea inside, and the actual size of the pea can be determined only by opening the pod. It is used primarily as a cooking vegetable.

In addition to the regular green (shell) peas, there are three types of edible-pod peas: snow, snap, and sugar snap peas. Peas are good both raw and cooked and are an excellent source of vitamin C. One three and one-half ounce serving of regular peas contains about 45 calories. Peas are used in salads, stir-fries, stews, soups, meat dishes, as a vegetable dish, with other vegetables, and as a garnish.

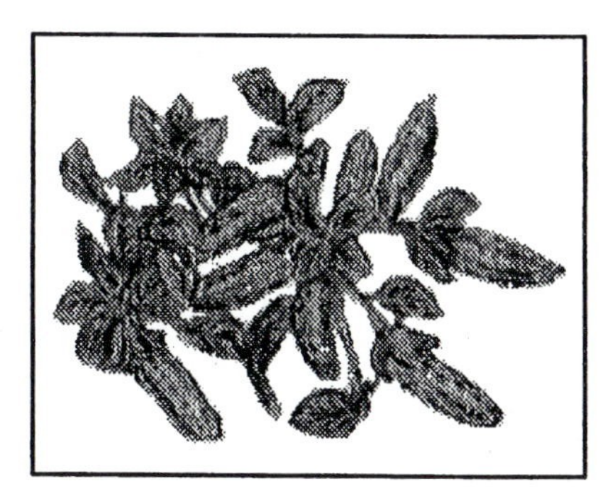

Pea Asparagus: A small creeping pea with red flowers and winged pods, the asparagus pea is native to the Mediterranean area where it is grows in fields and scrub. The winged pods are delicious if eaten young. Not grown commercially for the U.S. market.

Pea Pigeon: It is also known as Red Gran, no-eye pea, or dahl. Most commonly grown in India, it is an important crop in the Caribbean, Uganda, Malawi, and Southeast Asia. The peas may be eaten green, but most are used in the form of split peas eaten as Dahl. Available in dried form year-round.

Pea Snow: It is also known as sugar pea and China pea. Snow peas have flat pods that cling tightly to immature-appearing peas inside. This edible-pod pea can be eaten whole, cooked or fresh. Available from May through September.

Pea Snap: Snap pea have the appearance of the regular pea but smaller (about 40 percent smaller); they must be destringed before preparation. Snap peas can be eaten fresh or cooked. Available year-round.

Pea Sugar Snap®: The Sugar Snap pea is a cross between the snow pea and the green pea. Developed in 1979 by Calvin Lamborn, Sugar Snap peas are now the darling of the fresh pea industry. Tender and green, the plump pod is totally edible. They are used fresh in salads and Oriental cooking. Available year-round.

PEPPERS

Pepper Bell: It is native to Central America and Mexico. Discovered by Columbus in the West Indies and carried back to Europe, these long spicy vegetables quickly caught on, then spread to Africa and Asia, where they were incorporated into cuisine of countries all over the world. They were named peppers by accident when Columbus and his crew assumed them to be a fiery form of true pepper. Today, hundreds of pepper varieties are grown worldwide. Some are sweet and some have pungent, spicy flavors varying in degree from mild to extremely hot, and are generally referred to as chile peppers.

Sweet bell peppers will mature to various colors depending on variety, and as they mature, their sugar content increases. Most bell peppers are sold at an immature, green stage. The most common variety of bell pepper nationwide is the California Wonder, a square pepper that usually measures four to five inches in length with four lobes. The other sweet peppers most likely to be found in produce markets include red, yellow, purple, orange, white, and brown along with the tapered red pimentos, Hungarian sweet yellow wax peppers, and the slender, thin-walled Italian frying peppers. Sweet peppers provide delicious main courses stuffed with meat, rice, or seafood. When sliced in rings they make an excellent garnish, and their crisp texture and pleasant flavor make them a welcome addition to salads, stir-fries, soups, and stews. They can also be boiled or french fried as a separate vegetable. All peppers are high in vitamin C and are a fair source of several other nutrients. Sweet bell peppers are available year-round.

Pepper Le Rouge Royal®: There are two types of this particular pepper–Le Rouge Royale and Le Juane Royale, with rouge identified as red and juane as yellow. They are an uncharacteristically large pepper with a sweet, mild taste and brilliant flame-red or yellow color. A thick-walled pepper that grows up to ten inches long and four inches in diameter, it may weigh up to two pounds. It can be eaten fresh or cooked. Available from California mid-May through October.

Pepper Chile: Native to Central America and Mexico, these have been known and used for centuries as a flavoring and natural preservative for foods in Mexico, Central and South America, and the Caribbean. Chile peppers differ both in flavor and heat, depending on variety and type. Chilies of the same plant can range from mild to hot. Generally speaking, the smaller the pepper, the hotter it is. One of the world's hottest chile peppers, the habanero, is rated 100 times hotter than the jalapeno. Chile peppers are low in calories and an excellent source of vitamins C and A, potassium, folic acid, and some fiber. Most commercially grown chile peppers are available year-round. There are thousands of varieties grown, scores of which are produced in the United States.

Using the Scoville heat units, chile peppers range from 300,000 to 400,000 units for Habaneros to a mere 500 to 700 units for the El Paso chile. These pungency levels increase approximately tenfold when the peppers are dehydrated. There is considerable confusion in the names of chile peppers. In Mexico, a similar name may be applied to several different varieties of chile peppers, or the same chile pepper will have a different name for each locale, even though the growing areas may be less than 100 miles apart. Also, at times, U.S. importers will use tradenames instead of common names, confusing the issue even further. Some of the more popular commercial varieties follow.

Pepper Anaheim Chile: It is also known as long green, California long green, New Mexico chile, Rio Grande, chile verde, and chile colorado. Developed independently in California, the same strain in New Mexico is called chile verde when green and chile colorado when red. This is one of the peppers that, when dried, is a component of the graceful, deep-red colored strings or wreaths of peppers called ristras. Anaheims, named after a city in California where a pepper cannery opened, are mild (although they can be sharp) flavored peppers that are one of the most often used. Fresh, it is long, green, and tapering, with pods averaging six to seven inches in length and one and one-quarter to one and three-quarter inches in diameter. Anaheims are good stuffed with meat or cheese or chopped and added to salad dressings, omelets, tomato-based dishes, baked potatoes, or cornbread batter. California-grown Anaheims are usually milder than those grown in New Mexico. Heat range is from 1,000 to 1,500 units. Available year-round from California and New Mexico.

Pepper Ancho Chile: It is also known as pasilla. The most commonly used dried pepper in Mexico, it has a broad, triangular shape (approximately four and one-half inches long and three inches in diameter), a deep mahogany color, and a flavor that is hot. This chile must be soaked and ground before using as a sauce ingredient. Red chile sauce is its most common use although anchos are also used in making adobo, a marinade of chiles, vinegar, garlic, and oregano. Heat range is 2,500 to 3,000 units. Available year-round from California and Mexico.

Pepper Cherry Chile: There are two types of cherry peppers. One is hot and is used for seasoning or in sauces; the other is semisweet and used for pickling, sauces, and Mexican recipes. The heat range for the cherry chili pepper usually runs from 2,500 to 5,000 units. Available year-round from California, Mexico, and Texas.

Pepper Chile de Arbol Chile: Also know as bird's beak, the name of this cayenne-type pepper means "tree chile" in Spanish. A long, thin brilliant-red pepper, approximately three inches long and one-half inch in diameter, de arbols have a skin that is smooth, translucent, and rather brittle. It is comparable to the cayenne pepper and is popular in Cajun recipes. An orange-red pepper, approximately three inches long and one-half inch in diameter, de Arbols have a skin that is smooth, translucent, and rather brittle. This chile is very hot and is usually used for table sauce. Heat range is from 15,000 to 30,000 units. Available from Mexico year-round.

Pepper Fresno Chile: This fresh chile is named after Fresno, California. The Fresno has a small pod and medium-thick flesh that is light green in color like the Anaheim or cherry red when mature. It is as hot as the jalapeno and should be used for seasoning rather than a vegetable, in sauces (cooked or raw), or very sparingly in dips and salads. The chile is larger at its calyx end with a gradual tapering. It measures approximately one and one-half to two inches in length and about one inch in diameter. Heat range is from 3,500 to 4,500 units. Available year-round from California.

Pepper Habanero Chile: There are two popular commercial varieties: the hybrid Red Savina (the hottest) and the Scotch Bonnet. These are the hottest chile peppers in the world, 100 times hotter than the jalapeno. The skin is bright yellow-orange while the fruit is round and lantern-shaped. Highly aromatic with a fiery pungency, a few pieces go a long way. Used in Caribbean dishes and salsa. Heat ranges from 300,000 to 400,000 units. Available year-round from California and Mexico.

Pepper Hungarian Wax Chile: It is also known as sweet banana, banana pepper, and Hungarian yellow wax. It is a long, tapering, fairly narrow, translucent, waxy and creamy yellow (some become red-orange when mature) chile pepper. One form, usually called banana, has no heat at all but the other (Hungarian wax) is mild to moderately hot. Traditionally, these peppers are pickled as their thin skin does not require peeling. Hungarian peppers are good when slivered and used in fresh salads; they are particularly tasty in bean and grain dishes, uncooked sauces, dips, or relishes. Available year-round from California and Mexico.

Pepper Jalapeno Chile: It is also known as *en escabeche* in Mexico, originating in Mexico and named for the town of Jalapa in Veracruz. The jalapeno is a medium-hot pepper, usually no more than two inches long and three-quarters of an inch in diameter. It is long, pointed, and smooth-skinned, and is bright to dark-green in color, sometimes ranging to greenish-black. Jalapenos should be used sparingly to add spice to Mexican recipes, guacamole, salad dressings, egg dishes, and tomato-based dishes. Heat range is from 3,500 to 4,500 units. Available year-round from California.

Pepper Mexi-bell Chile: This pepper is a hybrid that was developed to create a mildly hot pepper that would not need peeling. The Mexi-bell can be red or green or a combination of both; it looks just like a bell pepper but smells like a chile. The heat can be controlled by keeping or removing the white-yellow membrane structure along the ribs. Uses include in Mexican dishes, eggs, salads, stir-fries, meat loaf, dip, and as a topping for pizza. Available mid-August through October from California.

Pepper Peperoni Chile: The peperoni is a small, red or green in color, sharp-edged, spearheaded, shiny, conical, and very hot chile used in Southeast Asian cooking. To subdue the fiery pungency, open the pod 24 hours prior to use and remove the inner membranes and seeds, drop into boiling water and let soak overnight. Used moderately, one teaspoon or less adds a fiery touch to soups, sauces, goulashes, stews, and tomato-based dishes. Heat range is from 6,000 to 17,000 units. Available year-round fresh or dried.

Pepper Poblano Ancho Chile: It is also known as pasilla. In Baja, California, it is labeled poblano; thus in California markets, two types of chiles use the same name. It is the fresh form of the ancho chile that is popular in the region of Puebla, its namesake. A long, tapered chile with a triangular shape, poblano is dark green, almost black in color, and mild to medium hot in flavor. It is used in meat dishes, sauces, and relishes. Heat range is from 2,500 to 3,000 units. Available year-round from Mexico and California.

Pepper Pasilla Chile: Also known as poblano, chilaca chile, or negro chile (when dried), this pepper variety is dark green, long, tapering, and narrow with a blunt end. This true pasilla is used in meat entrees, tamales, and quesadillas. Dried pasilla can be used for a thick, rich, dark sauce. Available year-round from Mexico and California.

Pepper Ristra Chile: This dried chile is a symbol of hope and plenty and usually comes dried in strings of various sizes and wreaths. Its history began along the Rio Grande in New Mexico and continues throughout many parts of Mexico and the American Southwest. These chiles are used as a preservative and flavoring for many vegetables, including such staples as potatoes, corn, and beans. The ristra is a long, dark brownish-red, wrinkled chile and is used to make red chile sauce. Heat range is from 35,000 to 40,000 units. Available year-round from Mexico.

Pepper Serrano Chile: One of the most widely used fresh chiles in Mexico and the Southwestern United States, its apparent origin is the mountain ridges (serranias) north of Puebla and Hidalgo in Mexico. Serranos are smaller (one to one and one-half inches in length) and more slender than the jalapeno, dark green in color, and about twice as hot. They are used in sauces, relishes, condiments, guacamole, and chili. Heat ranges from 7,000 to 25,000 units. Available year-round from California and Mexico.

Pepper Tabasco Chile: Tabasco peppers are tiny (about one inch in length), may be red, green, or yellow in color, slightly thicker than serranos, and about twice as hot. Used primarily in sauces. Heat ranges from 30,000 to 50,000 units. Available year-round from California and Mexico.

Pepper Yellow Chile: There are many varieties of yellow chiles available. They include yellow wax, Hungarian/Armenian wax, floral gem, bananas, and caribe. Taste testing is the only way to determine degree of hotness. Yellow chiles are used in salads, casseroles, meat stuffings, relishes, salsas, and Mexican dishes. Available year-round from California and Mexico.

Perilla: Perilla is a popular herb in Japan and is used as a garnish in the same way as parsley. The nettle-like leaves are bright green, reddish, or purple. Available at times in specialty produce stores.

POTATOES

Potato: The potato is the world's most important vegetable. Native to the Andean region of South America, it had its beginnings as a cultivated crop for the Native Indians. Spanish explorers brought the remarkable tuber back to Spain in the middle of the sixteenth century. From there, it spread to the rest of Europe. When first introduced in Europe, it was slow to be accepted because many people became sick from the toxicity that developed in green potatoes. Because of the presence of the alkaloid solanin, both the tomato-like fruit and the leaves are poisonous, as are any parts of the tubers that have turned green after exposure to light. Potatoes should always be stored away from light; even a small exposure can cause greening.

Because they were easily grown and grew well even in poor soil, the people of Ireland became dependent on them as their main food source. When a potato blight struck Ireland in 1845, 1846, and again in 1847, the great Irish famine began. Over 1 million people died of starvation or disease, and another 1 million Irish people migrated to the New World between 1847 and 1851 to escape this horror.

In 1872, Luther Burbank developed the now famous Burbank Russet potato, the most popular and the darling of the potato industry. This is the popular Idaho potato. Most potatoes grown in Idaho, Eastern Oregon, and Eastern Washington are considered dry-land potatoes (irrigated). Most potatoes grown in Western Oregon and Washington are considered wet-land potatoes (nonirrigated). A dry-land potato not only bakes well, but when whipped, will be light and fluffy. Wet-land potatoes, although they bake well, are usually more watery and less likely to whip to the creamy texture of a dry-land potato. Potato varieties are constantly changing with plant breeding. There are two classifications: baking and cooking varieties. These classifications reflect differences in baking and french frying qualities related to specific gravity. Those with a higher specific gravity are more mealy and better for baking, french frying, and mashing. Western United States produced (treated) russets and long whites that have a higher specific gravity than those grown in the Eastern United States. Potatoes run a gamut of colors, from shades of white, yellow, brown, red, to almost black-purple, depending on variety.

Potatoes kept in cold storage or in a cold room will convert some starch to sugar creating an even sweeter potato. By moving these potatoes to a warm, dry area, the sugar will usually convert back to starch. Care should be taken to keep potatoes from exposure to light as even a small amount of light will cause greening. An average eight ounce potato contains no fat, approximately 110 calories, and almost 50 percent of the MDA of vitamin C. The same size serving is low in sodium, high in dietary fiber, and provides 8 percent of the RDA of thiamin, iron, phosphorus, magnesium, copper, and folic acid and is a good source of potassium. Potatoes are available year-round. Some of the more popular varieties follow.

Potato California Long White: A large, long, smooth-skinned white potato that is good for baking.

Potato Chippewa: A large oblong potato that is good for cooking.

Potato Irish Cobbler: A white, roundish potato that is good for cooking.

Potato Katahdin: A round potato that is good for cooking.

Potato Kennebec: A smooth oblong potato that is good for chips.

Potato Livingstone: This unusual vegetable is used in many parts of Africa, from Nigeria to Transvaal. The tubers are elongated, about three inches in diameter, with irregular branching. When cooked they have the consistency and taste of potatoes and are used in a similar manner. Not grown commercially.

Potato Norgold Russet: A long potato that is good for french fries and baking.

Potato Red Pontiac: A large, red, oblong potato that is good for cooking.

Potato Russet Burbank: A long russet, good for french fries and baking.

Potato Sebago: A large, white, roundish potato that is generally good for cooking. Some of the other commercial varieties grown are the Norchip, Atlantic, Russet Norkotah, Centennial Russet, Shepody, Superior, Frito Lay, Hi Lite Russet, Frontier, Monona, and the new red Sangre.

Potato Yellow Finnish: It is also known as yellow flesh potato. This potato's seed originated in Peru from where it traveled to Finland. Now grown in the state of Washington by farmers who brought the seed over from their native Finland, this potato resembles a russet with brown skin and smooth texture. The flesh is pale yellow with a delicate, sweet butter taste. The care and use is the same as for other potatoes. Available from Washington State.

Potato Yukon Gold: A buttery-flavored gourmet potato discovered in Canada and later developed at Michigan State University. Averaging between one and two inches, the tiny potatoes are available from August to early March from Michigan.

PUMPKINS

Pumpkin: Like squashes, pumpkins are cucurbits and of North American origin. Certain hybrids grown for size are now approaching 1,000 pounds in weight, making pumpkins one of the largest vegetables grown in the world. The New England pie pumpkin types are used for cooking, but winter squashes provide better texture, flavor, and better yields per pound. This really is not deceptive, since there is no real distinction between the pumpkin and the winter squash.

Pumpkins in the United States are treated as an ornamental, with the small and medium orange varieties used as jack-o'-lanterns in celebration of Halloween. Besides the traditional use as a jack-o'-lantern, pumpkins are used in pies, soups, puddings, and stuffings. One three and one-half ounce portion of raw pumpkin contains 26 percent of the RDA of vitamin A and is a good source of potassium. Pumpkins are available late September through October. Four of the more popular varieties follow.

Pumpkin Connecticut Field: Noted for its shape, this variety is flattened on both the blossom end and the stem end. The pumpkin is thick fleshed with golden-orange skin and ranges from 16 to 20 pounds in weight. Commonly used as a jack-o'-lantern.

Pumpkin Howden: Developed by John Howden of Massachusetts, it was bred specifically for use as a jack-o'-lantern. The Howden has a round shape and firm structure; the rind is hard, ridged, deep orange in color, with dark green stems. With less tendency to produce flat-sided pumpkins, the Howden offers a more uniform shape than other varieties. Average weight is between 18 to 25 pounds.

Pumpkin Mammouth Gold: Usually coming from areas where the Howden are not grown, the Mammouth Gold is a nearly round pumpkin with a smooth, faintly ribbed shell. Its skin is golden-orange, and the average weight is 20 to 25 pounds.

Pumpkin Mini: It is also known as Jack Be Little or munchkin. This miniature pumpkin (squash) is good for both eating and ornamental purposes. The flesh has a fine grain texture and a mild sweet flavor, and the rind has a deep orange color. They average about two inches in height, three inches in diameter, and weigh between eight ounces and two pounds. Available September through October from California.

R

Radicchio: It is also known as red chicory or red-leafed chicory. This red-leafed vegetable has been cultivated in Italy since the sixteenth century. This round, deep burgundy to red-bronze chicory is shaped like a cabbage rose with close-wrapped shiny, smooth leaves with white central ribs. It resembles a sharp-petal tulip and may be as small as baby endive or as large as a small romaine. Radicchio has a distinct bittersweet flavor that is similar to that of escarole and Belgian endive. Radicchio is an unpredictable grower. It begins life as bright green leaves, then turns deeper green, then red as the weather cools. Some, however, never turn red and may even change the shape of both head and leaves as the plant matures. Uses include in salads, as a garnish, and in Italian dishes. Available year-round from California.

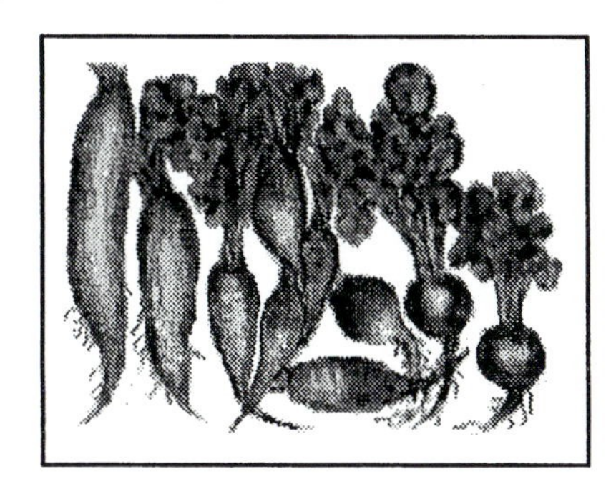

Radish: The radish is native to Southeast Asia and China. Quick-growing radishes get their name from the Greek word for fast-appearing. Cultivation is traceable to ancient China and Egypt. This member of the Cruciferae family (of which mustard and cabbage are subfamilies) comes in many varieties of different colors (mostly variations of reds, whites, and blacks), sizes (up to 60 pounds each), and shapes (as long as four feet). A variety called Easter Egg even produces several different colors. The giant of the family is sakurajima, a Japanese radish that can reach a diameter of 32 inches and a depth of five inches. Another Oriental radish, moriguchi can reach a length of four feet and may be the world's longest. The small, round red varieties and the small, white, carrot-shaped varieties are preferred in America. Radishes are not exciting in terms of nutritional benefits, but they add color and crispness to salads, slaws, and sandwiches, as well as adding their pleasantly sharp taste. Popular in relish tray assortments and for decorative carving (usually roses), radishes are thought to be a good appetite stimulator. Available year-round.

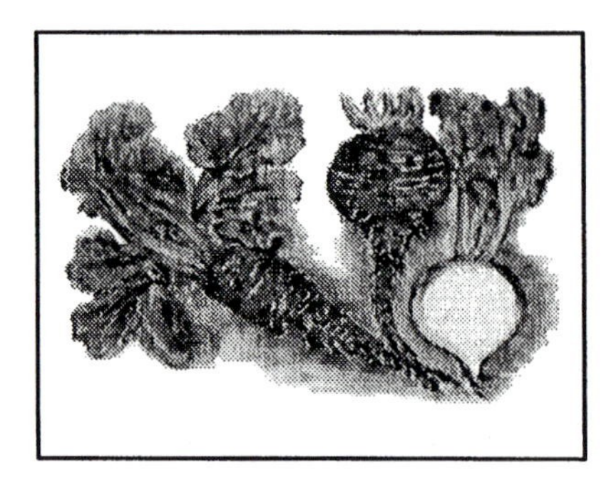

Radish Black: Native to Asia and Russia, black radishes are almost as pungent as horseradish. Firm and rather dry, they are nothing like the petite, juicy, crisp, scarlet-red, familiar round radish. Having the shape and size of turnips (about two to six inches in diameter), sooty black or black-brown in color, they have firm white flesh and must be peeled before eating. Black radishes taste like firm, rather strong turnips when cooked. They can be used fresh in salads, creamed, or cooked like turnips. Available year-round.

Rakkyo: Also known as Jiao tou and Tsung tao, this plant is grown in China and Japan for its bulbs, which are like small shallots. The plant is perennial, forming dense clusters of narrow bulbs. Rakkyo is used in Chinese cooking either raw or cooked like onions, but is more popular as a pickling onion. Available year-round.

Rape: It is also known as colza. Rape is another member of the cabbage family and is grown for forage and for the extraction of oil from its seeds. It is used much like kale and collard, as a cooking vegetable, for its flavor and excellent nutritional benefits. Available sometimes in specialty produce departments.

Rapini: It is also known as broccoli raab and Chinese flowering cabbage. Rapini has dark green, chard-like leaves on a stalk, and a slightly bitter flavor. Popular in Italian and Chinese cooking, rapini can add zest to potato or pasta dishes. Rapini can be cooked like broccoli, but requires less cooking time. Rapini can be boiled, braised, stir-fried, sautéd, or steamed. It contains calcium and iron and has about 40 calories per cup. Available August through March from California.

Rhubarb Cherry: Little is known about the history of rhubarb except that it probably originated in Tibet, spreading to Europe about 300 years ago. An enterprising Maine farmer is credited with bringing it to the New World in about 1800 where it became popular for tarts and pies.

The plant has a huge rhizome (stem below the ground) that produces thick, long, fleshy leaf stalks in the spring. Commercial producers sometimes put rhizomes in forcing beds in hothouses where light can be limited to induce long stalk formation. Only the stalks are eaten. Roots and leaves contain substances that cause severe illness, even death. Of the 100-plus varieties, Victoria and Linnaeus are the old standbys. Ruby, McDonald, Cherry Red, and Valentine have been developed to increase the amount of red color in the stalks. Cut stalks are boiled and stewed, often with raisins, and served like a breakfast fruit or made into pies and marmalades. Combined with strawberries, rhubarb is one of the most popular pies in New England. Rhubarb is promoted for its unique and zesty flavor when cooked. Available March through May from Washington.

Rhubarb Strawberry: Strawberry rhubarb differs from cherry rhubarb in that it is grown in greenhouses and has a much smoother and more delicate texture when cooked. Its uses are the same as cherry rhubarb. Available from January through mid-April from Washington.

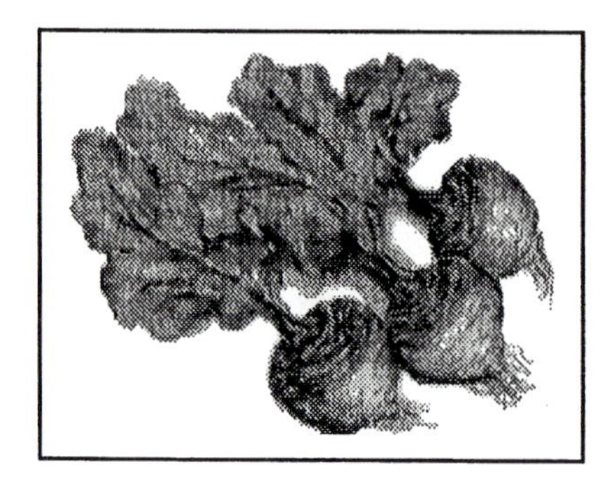

Rutabaga: Also known as Swede or Swedish turnip, it is thought to be native to northern Europe. The name comes from Sweden and the vegetable probably resulted from a chance hybridization, centuries ago, between cabbage and turnips. The plant is part of the cabbage family and cousin to the turnip. Most varieties are yellow fleshed. Rutabagas are the largest of the root vegetables, round, and with a distinct turnip shape of which they are often confused. Their uses include being boiled, mashed, used in stews and soups, and eaten raw. They provide a large amount of vitamin C and potassium and are low in calories (before they are buttered). Available year-round from Canada.

S

Salsify: It is also called oyster plant, vegetable oyster, or white salsify. Salsify is a word used to describe two plants (white oyster plant and black oyster plant), both of which belong to the vast Compositae family. The salsify root has been cultivated for about 400 years. Salsify is a hairy, white-rooted plant that is peeled, sliced, and cooked like parsnips. The root vegetable itself is shaped like an irregular parsnip with tiny rootlets, has a thin, pale tan skin, and when cooked really does have an oyster-like flavor, though it is very mild and delicate. Used in soups, stews, and braises, salsify offers calcium and iron as well as 35 calories per one-half cup. Available from October through March.

Scorzonera: This unusual vegetable is also known as black salsify, black oyster plant, or viper grass. The name comes from either *escorzo nera*' Spanish for black bark, or from *escorco*, the Catalan word for viper, so-called because the plant's juice was once thought to be an antidote to snakebite. This edible root resembles a muddy brown, nontapering carrot and is smoother, longer, and more regular in shape than white salsify. Both white and black salsify are delicate in flavor. When peeled, salsify darkens almost immediately and should be placed in water with a little lemon juice to preserve the color. Its use is the same as white salsify. Available from October through March.

Seakale: Also known as Chou marin, it is native to Europe, England, and eastern Asia. Seakale is a long-lived perennial, growing wild along the Atlantic Coast and the shores of the Baltic and Black Seas. The plant is blanched in early spring, when the leaves and flowering shoots have a mild flavor. They are usually cooked like asparagus and served with butter or a mild sauce. The young flowering shoots may also be eaten like sprouting broccoli. Not grown commercially.

Shallot: Shallots are believed to be native to Asia Minor. They are a member of the Amaryllis family and are cultivated more in England, Europe, Africa, and Asia than in the United States, where the name shallot is applied to a small red onion. The shallot is known as the "queen of the sauce onions." Its leaves are slender and cylindrical, and the pear-shaped bulb splits into bulblets (cloves). The outer skin is gray at the top and red below, and the flavor is milder and more aromatic than that of other onions. The cloves are minced to flavor soups, stews, and meats. They are also pickled and the young leaves are eaten in salads. A 3.5 ounce serving contains about 72 calories. Available year-round.

Shepherd's Purse: A common weed found throughout the world, the plants are annuals forming rosettes of jagged leaves that may grow to eight inches in good soil. It is used raw in salads, cooked like spinach, or in the Chinese style, floating in a clear soup. Available at times from California.

Skirret: This plant is native to central Europe, Siberia, and central Asia. Skirret is an ancient vegetable, now seldom seen, and its origin is unknown. It is a perennial, producing a cluster of roots about three inches in diameter. They are said to taste sweet and floury and can be used in much the same way as parsnips. Not grown commercially.

Sorrel: It is also known as French sorrel or sour grass. A relative of the rhubarb family, sorrels grow wild in northern Europe and Asia and are used extensively by the French, Norwegians, and Laplanders. Sorrels are small, low-growing, leafy green plants used for soups, in combination with other greens in salads, with fish, soups, sauces, and custards, both sweet and savory. The plant produces a fairly tight bouquet of small leaves (up to one inch long), some pointed and others rounded and oval. The high content of oxalic acid makes the plant sour, so it is used sparingly as a seasoning or the flavor is diluted by mixing with other vegetables. Available year-round.

SPINACH

Spinach: Native to Persia, it is one of the best known vegetables. Spinach was cultivated in Persia more than 2,000 years ago. The plant is closely related to beets and Swiss chard. There are a number of spinach varieties and they fall into two broad groups: smooth-textured types and those having ruffled or crinkled leaves. There is no difference in flavor or nutrient content. Unfortunately, spinach is thought of in most households as a cooked vegetable, but it is excellent when used in soups, in combination with eggs or bacon, in quiche, or used raw in salads. Spinach is an excellent source of vitamin A and potassium and a good source of vitamin C, riboflavin, and iron. Twelve ounces of spinach contain about 90 calories. Available year-round from California.

Spinach New Zealand: A coarser-leafed relative of regular spinach, New Zealand spinach has a mellow taste when cooked and a flavor that is similar to grass with a tang when raw. The leaves are dark green with a slight fuzz. Best when cooked, this spinach can be added raw to salads. Nutritional value is about the same as for regular spinach. Available year-round from California.

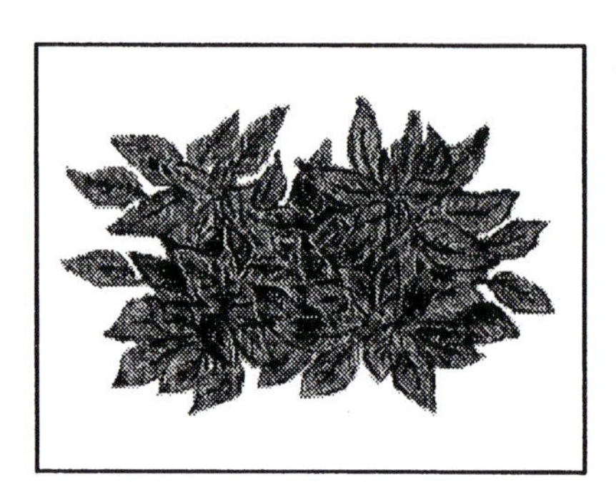

Spinach Chinese: It is also known as Tampala spinach, Yin tsoi, and Bayam. This plant, commonly grown in India, Southeast Asia, China, and Japan, is used as a substitute for spinach. The whole plant is edible, but the young shoots and leaves are usually eaten. Available from California.

Spinach Water: Also known as kancon and green engstsai, it is thought to be native to Southern India. Water spinach is an aquatic plant with edible young shoots and leaves that can be served and eaten like spinach. In many warm areas such as Florida, it has become a serious weed. Available from Florida.

SPROUTS

Sprouts: Alfalfa and mung bean sprouts are the most popular and best known sprouts, but any number of seeds and beans can be sprouted for a crunchy addition to salads, sandwiches, and other dishes. Some sprouts like the tiny alfalfa and the larger mung bean are mild in flavor while others, such as mustard, cress, radish, and chia are more peppery. While most sprouts are eaten raw or used as a garnish, mung beans and soybeans are a common ingredient in cooked Asian dishes. Sprouts have a high ratio of proteins to calories. Most sprouts are available year-round. Some of the more popular sprouts follow.

Sprout Alfalfa: Alfalfa sprouts differ from other sprouts in that they do not have a specific seed producing a specific variety. Until recently, there have been no seeds grown solely for sprouting purposes. The seed has many characteristics and must be the purest available, highest in germination count, lowest in extraneous crop seed, and lowest in weed seed. Alfalfa sprouts are white with tiny green tops and offer a subtle nut-like flavor and crisp texture. They should be used before they reach a height of five inches.

Sprout Bean: These can be grown from a variety of seeds, all resulting in a product with roughly the same taste and nutritional value. Sprouts from mung beans can be grown from red, yellow, green, or black seeds. Mung beans, when germinated, become quite fragile, the growing cycle is rapid, and the sprouts are ready for use in just three days after the first soaking. Mung beans are a staple in Oriental dishes.

Sprout Clover-Radish: Developed after three years of research, these spicy sprouts resemble alfalfa sprouts. Tangy and crisp, they should be kept refrigerated at all times. Available year-round from California.

Sprout Daikon-Radish: It is also known as Kaiware. Daikon-radish sprouts are similar to radishes. Small sprouts with a hot flavor, they are grown hydroponically and they add zest to many dishes such as salads, sandwiches, sushi, and tempura. Available year-round from California.

Sprout Onion: Onion sprouts offer another opportunity to get a pleasant taste of onion. These delicate little sprouts can be used on hamburgers, tacos, and salads and as a complement to avocados. Onion sprouts are handled in the same manner as alfalfa sprouts. Available year-round from California.

Sprout Stir-Fry: A combination of nutty-flavored sprouts of mung, azuki, and lentil beans. An excellent addition to stir-fries, salads, soups, and steamed vegetables. Available year-round from California.

Sprout Sunflower: Produced from sunflower seeds, these have a mild sweet flavor that is similar to alfalfa sprouts. They are very crunchy. Sunflower sprouts are grown like alfalfa sprouts, but for a longer period in order to produce a larger sprout. Excellent in salads and sandwiches. Available year-round from California.

Sprout Soybean: Soybean sprouts are larger than mung bean sprouts and have a stronger flavor due to the greater amount of chlorophyll. They are used in Oriental cooking, stir-fries, salads, and soups. Available year-round from California. Other types of sprouts available are garlic, dill, pumpkin, wheat, and lentil.

SQUASH

Squash: Native to North America, the word squash came from the Algonguin Indian word *askutasquash*, meaning "eaten raw." Squash was unknown in Europe until early explorers returned from America with the seeds. Squash was an important Indian food; however, few shared the desire to eat squash raw until the past two decades when raw summer squash types started to be used in salads.

Summer squash are high in vitamin C and very low in calories. A cup of cooked summer squash contains about 25 calories, while cooked winter squash contains about 90 to 130 calories per cup. Winter squash are low in sodium and an excellent source of vitamin A. All squash are high in fiber.

Squash comes in an array of sizes, shapes, and colors. They are classified as either summer (soft-shelled) or winter (hard-shelled). Winter squash will keep for extended periods, usually from one to six months, depending on the variety. Summer squash is usually picked immature while tender and used immediately or within four days of harvest; under optimum conditions, storage life can be extended up to two weeks. The summer squash (harvested immature) are divided into three subtypes: Neck types (smooth and warted) include yellow straightneck and golden crookneck; Scallop types include (white, yellow, or light green) white bush and patty pan; and Italian types (vegetable marrows) are zucchini (long green or striped), golden zucchini (long yellow), and cocozelle (long, dark green).

Winter squash (harvested mature) are divided into two subtypes: small and large. Some of the small winter types are the Danish, acorn, or tablequeen (dark green, ribbed, oval, with pointed blossom end, one to three pounds); golden acorn (new orange, yellow, white colored acorn types, one to three pounds); butternut (bell-shaped, tan, with thick neck and bell-shaped blossom end, two to four pounds); delicata (oblong or club-shaped, tan with green stripes, one to two pounds); buttercup (green with turban cap over blossom end, three to five pounds); turban (coarse, reddish-orange with turban cap, four to eight pounds); golden nugget (round-shaped, salmon-colored, one to three pounds, with a large seed cavity and a very mild flavor).

Some of the large winter types are: banana (long, cylindrical, often pinkish in skin color, up to 150 pounds); Hubbard (dark green, ribbed, rough bumpy skin, pointed at both ends, up to 80 pounds; also Blue Hubbards, which are blue-gray in color); marblehead (pale bluish-gray skin, smooth, rounded, up to 80 pounds); Golden Hubbard (salmon to orange in color, slightly smaller than Green Hubbard, up to 60 pounds); delicious (green or golden in color, shaped like a top, 5 to 15 pounds); orange or Boston marrow (orange skin, roundish, 12 to 16 pounds).

Squash Australian Blue: Once available only in Australia, this squash is now grown in California. The blue-gray shell reveals a thick, orange flesh that is soft and mild as a pumpkin. Averaging eight to ten pounds in weight, the Australian blue squash is used for baking, desserts, soups, scones, and as a vegetable. Experts say it seals itself naturally and bad spots can be cut away without spoiling the rest of the squash. Available year-round from California.

Squash Calabaza: Also known as Cuban squash, toadback, zapallo, or West Indian pumpkin, it is native to the Americas. A common hard-shelled squash in the vegetable markets of Central and South America and the Caribbean, it is usually marketed in chunks. Although usually large (up to 100 pounds), the calabaza may be as small as a honeydew, round or slightly pear-shaped with fine-grained, sweet, moist (not watery), bright orange flesh. The mottled skin may be forest green, sunset, buff, speckled, or striated and is fairly smooth. Calabaza is used in soups, salads, stews, purees, sauces, cakes, pies, custards, quick bread, cookies, and puddings–in just about any recipe that calls for pumpkin. It is very low in calories, about 35 calories in a cooked one cup serving, rich in vitamin A, potassium, and folic acid and very low in sodium. Available August through March.

Squash Chayote: It is also known as christophene, vegetable pear, chocho, choko, pepinella, xuxu, and mirliton. Chayote squash has been cultivated for centuries by the Aztecs and Mayans, and it is a member of the gourd family. Commercially grown, it is similar in size and shape to a large pear, with light to dark green skin, usually ribbed, and with the taste and consistency of a cucumber mixed with apple and zucchini. Popular throughout Mexico, Latin America, the Caribbean, and in Creole cooking, it is mostly baked, fried, or boiled. Chayote is a good source of potassium and fiber and is very low in sodium. A cup serving contains about 40 calories. Available year-round.

Squash Delicata: This small squash is also known as sweet potato squash. The delicata squash has been around since 1894 when it was introduced by the Peter Henderson Company (now defunct) of New York City. It is lightly ridged, oblong, six to nine inches in length, two to three inches in diameter, and may weigh as little as six ounces (rare in hard-skinned squash) or up to three pounds. The skin is ivory, mottled and striped with green (occasionally with orange), and the yellow flesh tastes like a blend of butternut squash and sweet potato. Fine, moist, and creamy, the meat is of excellent quality, and the skin, properly cooked, can be tender enough to eat. Delicata is a good source of vitamins A and C, iron, and potassium and is low in calories and sodium. Available July through November.

Squash Kabocha: Kabocha is a generic grouping as well as a marketing name in this country for many strains of Japanese pumpkin and winter squashes. These Oriental squash, of flattened drum or turban shape, with deep green skinned, resemble the buttercup squash. Kabochas range from one to seven pounds, but average about two to three pounds. Their mottled rind is thick, deep green with paler, uneven stripes and markings. The cooked yellow-orange, fine-grained flesh is tender and floury dry with a flavor balanced between sweet potato and pumpkin. It can be baked, steamed, pureed, braised, or chunked. Like all winter squash, they are rich in vitamin A and have good amounts of vitamin C, iron, and potassium. Available year-round from California.

Squash Mushroom: Similar to zucchini, the white, mushroom-shaped squash grows in tropical areas year-round. It comes in a variety of shapes and sizes. The squash should be young and eaten with skin and seeds. To prepare, cut washed vegetable into good-sized chunks, fry with chopped onion, garlic, and bacon, season with herbs, and serve. Available year-round from California.

Squash Sweet Dumpling: Developed about 25 years ago by the same seed company (Sakata) that brought us the kabocha and spaghetti squashes, this squash is dubbed "vegetable gourd" in the company's catalog. Looking like a small-tufted hassock, this small, rounded squash usually weighs between 8 and 16 ounces. It is cream colored and scalloped with regular, shallow lobes, and its indented areas are striped in mottled ivy-green. The flesh is pale yellow, smooth-textured, mildly sweet with a slight corn flavor, and dry like a potato. It can be baked or steamed and the smaller squash are appealing as individual servings. Available from California.

Squash Spaghetti: No one is really sure where this squash came from. Frieda Caplan of Frieda's Finest Produce Specialties, Inc., a Los Angeles based distributor of fruits and vegetables, was the leader in introducing it to the American consumer. The spaghetti squash is a hard, cream-to-yellow, oval vegetable shaped like a cross between a football and a small pumpkin. When cooked, the flesh resembles long strands of spaghetti. People who love this squash seem fascinated by its novelty. There is no doubt that it is unusual to cook a vegetable in the shape of a watermelon, open it up, and pull out what appears to be miles of crisp-tender, golden strands. Usually baked and served as you would spaghetti with a full-flavored but not overpowering sauce. Spaghetti squash is low in calories, an excellent source of folic acid, quite high in fiber, and has a fair amount of potassium, vitamin A, and niacin. Available year-round from California, Texas, and Mexico.

Squash Blossoms: Although these natural appurtenances precede every squash that is grown, they are usually difficult to find. Golden-yellow and fragile, they are picked just before opening and must be used shortly thereafter. At their simplest, cleaned, sliced, and sautéed, squash flowers make a beautiful garnish for steamed, mashed zucchini, fried tomatoes, thick creamy soups, or vegetable purees. Squash flowers are very low in calories and they contain vitamin A and calcium. They are high in fiber and low in sodium. Not grown commercially.

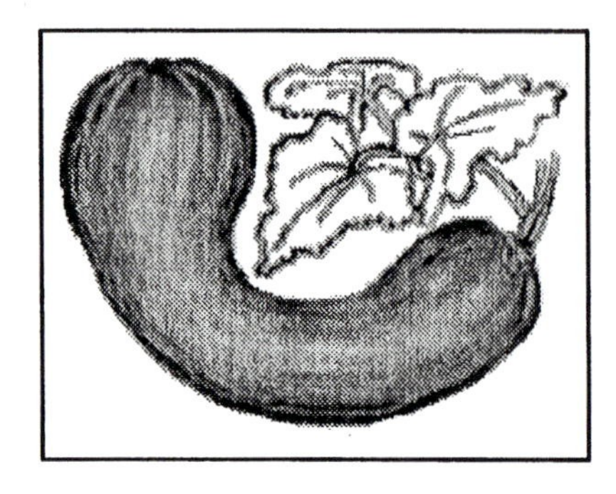

Squash Tahitian: New to the United States, the Tahitian resembles butternut squash with its large crescent shape. Deep green, the squash turns golden yellow when ripe, although it can be picked green and later ripened indoors. Flavor and fragrance is more like a sweet melon or mild sweet potato, and it can be eaten raw like a melon or cooked like a squash. The deep orange flesh has a high sugar content and when frozen it still retains its flavor after several months. Available year-round from California.

Swamp Cabbage: Also known as hearts of palm, it is the leaf-producing bud of the cabbage palm. The leaf-producing bud of this palm is harvested as an edible delicacy. The taste of the bud is similar to raw cabbage, hence the common name of the tree, cabbage palm. In the early days of Florida, picnickers could go to the riverside with a fishing pole, a frying pan, and an ax, and lunch would be freshly cooked fish and swamp cabbage salad. Fresh swamp cabbage or hearts-of-palm has become hard to find in present-day Florida. Not grown commercially.

Sweet Chestnut: It is also known as marrons. Sweet chestnuts are the only nut treated as a vegetable. Because they contain more starch and less oil than other nuts, they can be cooked in a variety of ways. Use in soups, cereals, stews, and stuffings or eat whole, roasted, steamed, or boiled. Once shelled, sweet chestnuts are preserved whole in sugar syrup as marron glazes. Available November through January.

Sweet Potato: Also known as batata and boniata, it is native to Central America and a member of the Solanaceae family, which includes the subfamilies nightshade and morning glory. The sweet potato as known in the United States is a white variety grown in New Jersey, Maryland, Virginia, and California. The white or yellow Jersey is a dry type of sweet potato whereas the Louisiana yam (sweet potato) is more moist, and its skin is a coppery color as opposed to the yellowish hue of the white variety. The baked white potato has only a trace of vitamin A while the baked sweet potato has almost 9,000 international units of that important vitamin. Available year-round.

Compared to a baked white potato in its skin, sweet potato contains four times as much calcium, 25 percent more iron, approximately the same level of vitamin C, and only 47 more calories.

T

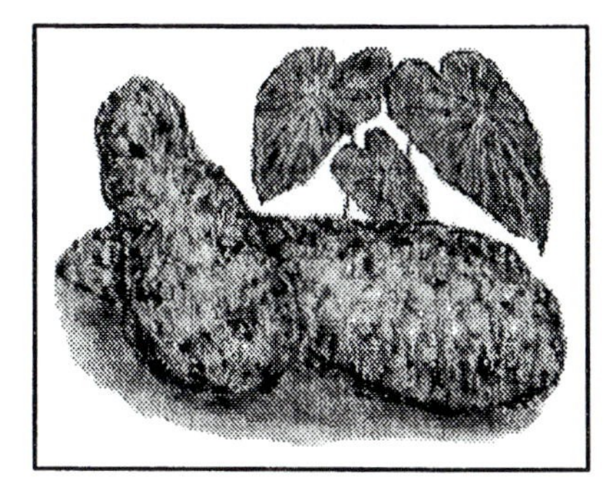

Taro: Also known as dasheen, malanga, tannia, eddo, coco, Woo tau, and Hung nga, it is thought to have originated in the East Indies. Taro has two distinct forms, but the most common is one but ranges from one to three pounds in size and is the key ingredient in Hawaiian poi.

It is brown, shaggy, cylindrical in shape, circled with distinct rings, and small rootlets usually twist from one end. The flesh is smooth, very white, cream or lilac-gray in color, and sometimes speckled with brown. Its flavor and texture taste like a combination of potato and chestnut.

The other form is small, about the size of a little new potato, and elongated. Shaped like a kidney bean, it has a much smoother form, is quite hairy, and may sprout a pinkish bud at the tip. This cormel, which attaches laterally to the larger corm (described above), is favored in China and Japan, where it is called eddo. Taro root must be eaten cooked with skin removed. Use it as a vegetable or add to soups and stews. Less sticky than other tropical roots (such as yucca, yam, and malanga), taro can be cooked, then pureed with enrichments for croquettes, fritters, and soufflés. Taro is a good source of potassium, a fair source of fiber, low in sodium, and one-half cup contains about 56 calories. Available year-round.

Tomatillo: It is also known as Mexican husk tomato, jamberry, tomatito verde, fresadilla, miltomate, and Chinese lantern plant. It is native to Mexico and Central America. Cultivated since the Aztec times, tomatillo is the best known of about one hundred *Physalis* species, among which are the ground cherry and cape gooseberry. This group of fruits are enclosed in papery calyxes that cover them like Oriental lampshades. Dry as autumn leaves and parchment-colored, the web-like enclosure is easily peeled off to reveal the fruit. The tomatillo ranges from an inch in diameter to plum-sized, and when husked, resembles green cherry tomatos (both are members of the Solanacea family, which includes the subfamily nightshade), but is more lustrous and firm. It has solid, seedy flesh and a slight, sweet apple or plum flavor with lemon overtones that is enhanced by cooking. Although it is purplish in color and may ripen to yellow, it is commonly used green. Tomatillos are almost always cooked to develop their flavor. Traditional uses include steaming and adding to salsas, Mexican stews, and casseroles. They are a good source of vitamin A, vitamin C, and niacin, and a one-half cup serving contains about 50 calories. Although tomatillo is classified as a fruit, it is used as a vegetable, and most produce departments display it in the vegetable section. Available year-round from California and Mexico.

Tomato: Controversy has long raged over whether the tomato is a fruit or a vegetable, but legally it is a vegetable, so ruled by the U.S. Supreme Court in 1893. The ruling came because of the different trade regulations governing fruits and vegetables. The tomato (actually a berry) is native to the Americas. It was initially cultivated by Aztecs and Incas as early as 700 AD. The name tomato is derived from the Mexican (Mayan) term *xtomatl.* Europeans first saw the tomato when the Spanish Conquistadors brought the seeds back from Mexico and Central America in the sixteenth century. For many years tomatoes were grown as ornamental plants, but once discovered as edible, they quickly found favor in Spain, Portugal, and Italy. As the tomato traveled north, it was veiled in mystery. The French called it "the apple of love," the Germans "the apple of paradise." The British, while admiring its brilliant red color, disclaimed it as a food, fearing it was poisonous, probably because the immature small green berries resembled the fruit of its relative, the poisonous nightshade plant. The same fear persisted among the colonists in the United States until the early nineteenth century, but in the early 1800s, the Creoles in New Orleans put their cooking on the map with their tomato-enhanced gumbos and jambalayas. The people in Maine quickly followed suit, combining fresh tomatoes with local seafood. By 1850, tomatoes were an important produce item in every American city.

Tomatoes range in size from the popular cherry tomato (about the size of a large marble) to one and two pound giants (beef steaks). Most of the tomatoes we are familiar with are round and red, but there are yellow, orange, and even pink ones. Pear-shaped, Roma types are very meaty and good for canning and sauces; the rounder and cherry types are used most often raw in salads, dressings, stews, sandwiches, baked dishes, and scrambled eggs, to name only a few of their uses. A medium-size tomato (about 150 grams) contains about 35 calories, 40 percent of the RDA of vitamin C, and 20 percent of the vitamin A allowance. They are cholesterol free, very low in sodium (only 10 milligrams), and contain phosphorous, essential folic acid, copper, and other minerals. Tomatoes should never be refrigerated until fully ripe, then only for a few days, since storing longer results in cell breakdown and flavor deterioration. Today, California and Florida are the leading domestic producers. Tomatoes are available year-round.

Turnip: Native to northern Europe, turnips have been used for centuries as both food and medicine. This member of the cabbage family was once regarded as food for the poor and common people–an unfortunate degradation because properly prepared, turnips are an uncommonly fine-tasting vegetable, and a good source of vitamin C with other nutrients as well. The globe-shaped roots have reddish-purple skin, crisp, white flesh (unlike the yellow-fleshed rutabagas), and a sweet-to-hot flavor. They are excellent in soups, stews, raw in salads, in boiled dinners, casseroles, and as a separate vegetable. An Italian turnip with a small root that is cooked with the tops and roots together is known as rappini (pronounced rap-pean-e), and is usually grown for markets in Italian neighborhoods. Available year-round.

U

Ulluco: Another vegetable from South America, it is also known as papa lisa and melloca. Ulluco is an important crop of the high

Andes, along with potatoes, anu, and oca. The tubers are similar to small potatoes; they may be round or elongated with color ranging from yellow, yellow-magenta, to magenta-pink. The tubers are used much the same as potatoes and are used in a local Andean recipe called *lingli*. Not grown commercially in the United States.

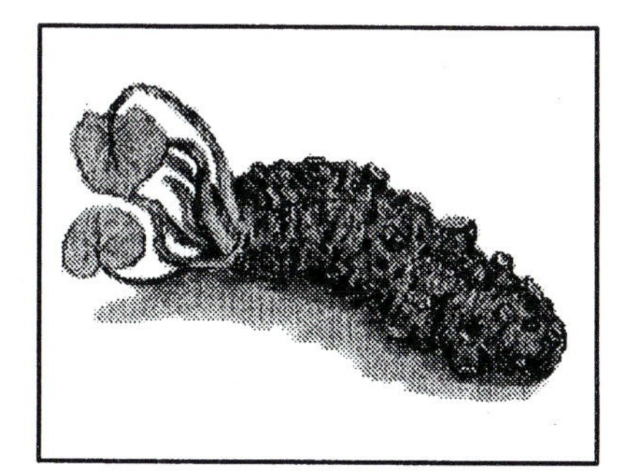

Wasabi: Wasabi is native to Japan and the Sakhalin Islands and a close relative of the true watercress family. It has a thick rhizome, with long-stalked, kidney-shaped, toothed leaves, and few-leafed flowering stems to 18 inches tall. Highly prized by the Japanese for its pungent flavor, this is the plant from which comes the green sap that forms the sauce for sushi. The rhizome is used to make the paste, which tastes similar to horseradish and goes well with raw fish such as tuna. Also, when wasabi is grated and mixed with soy sauce, it makes a great dip for grilled fish or meat dishes. It is also grated into soup just before serving. Available year round.

Water Caltrop: Also known as Ling kok and water chestnut (a misnomer), this plant (*Trapa bicornis*) differs from the original water chestnut, not only in the way the plant looks but in the shape of its corms. The usual corm has two large, recurved horns on the large shiny black fruit or seed. Water caltrop is starchy, similar to the English chestnut, and is used in much the same way as regular water chestnuts. Caltrops should be peeled before use, and the seeds must be boiled at least an hour prior to eating since they are poisonous when raw. Available year-round.

Water Chestnut: Water chestnuts are also known as Chinese water chestnuts, Mah tai, and Kuwai. There are two Asiatic aquatic plants that are known as water chestnuts. The one here (*Eleocharis dulcis*) is considered the real Chinese water chestnut. This plant grows in the mud along the edges of lakes or ponds and in marshes or flooded fields where they are cultivated on a large scale by the Chinese. Planted corms send out underground root stems that develop rush-like tubular stalks, which rise to the surface with a circle of floating leaves, then small flowers. Water chestnuts in their fresh form look like muddy little tulip bulbs or chestnuts that are about one and one-half inches in diameter. The skin is brown-to-black, the flesh is crisp and white with a unique texture, and the flavor is similar to the true chestnut. Fresh water chestnuts are highly perishable and should be kept in plastic and refrigerated until they are used. Water chestnuts are used in Oriental cooking, stir-fries, in salads, vegetable, rice, noodle, seafood, and poultry dishes. They are low in calories at 35 calories per cup. Available year-round.

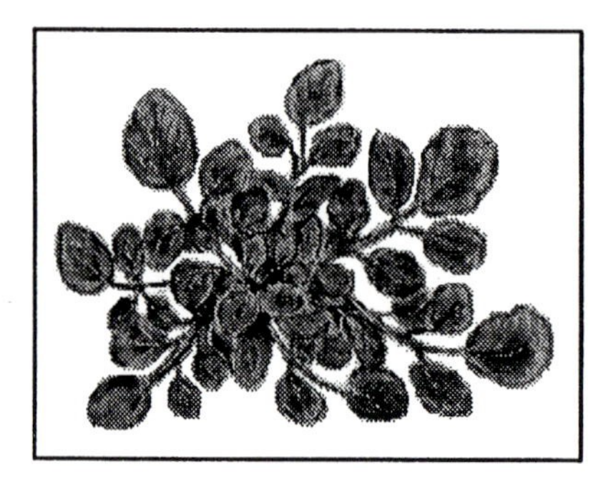

Watercress: This member of the mustard family grows best alongside or even partially submerged in shallow, running, fresh water, where it is cultivated in special beds in mid-Atlantic states and the South. This leafy plant has dime-sized, dark green leaves and eight- to ten-inch stems. Its spicy flavor makes it a popular garnish as well as a salad and sandwich ingredient. Watercress can be used like any lettuce or steamed with fish, vegetables, or poultry. It is a good source of vitamin A, and one-half cup chopped watercress has about two calories. Available year-round.

Winter-Cress: Winter-cress is an easily grown perennial, with rosettes of pinnately lobed, dark-green leaves. They are powerfully flavored and generally eaten in salads, although they are probably better in soups. Winter-cress has been grown as a salad crop for at least 300 years and is used as a substitute for watercress. Available from California.

In ancient lore, watercress was believed to be a cure for derangement, and if eaten before drinking alcoholic beverages, it would help keep one sober.

Winter Melon: It is also known as winter gourd, ash pumpkin, Chinese preserving melon, Kundur, and Dung gwa. Ironically, winter melon grows in the tropics and is harvested in the summer. The name comes from the waxy white blotching on the mature melon, which looks like a light dusting of snow and from the fact that it keeps well in cold storage during the winter months. Despite their resemblance, winter melons are not related to watermelon or other true melons, but are squashes or more properly, wax gourds, like the small, fuzzy melons

The most spectacular use of winter melon is as a serving vessel, sometimes intricately carved. Although an odd choice, since the fruit is not sweet, winter melon is a popular candy preserve. It is also the filling for "melon cakes," a flaky pastry sold in Chinese bakeries. Watermelon green in color, winter melons grow to 100 pounds or more, although those harvested around Sacramento, California are much smaller. Winter melon may be purchased by the pound or by the whole melon. It is usually used in combination with other ingredients including chicken, ham, mushrooms, abalone, and bamboo shoots. Available from California.

Y

Yacon: Yacon is native to Peru. It is a tall leafy plant with yellow daisy flowers that produces large, dahlia-like tubers that can be eaten raw. The taste is similar to a sweetish water chestnut. The flavor of the tubers are said to be improved after exposure to the sun. Like the Jerusalem artichoke, their sugar is inulin, not sucrose. Not grown commercially for the U.S. market.

Yam Chinese: This large yam, also known as mountain yam, Shu yu, and Dai shue, is an attractive, hardy, twining plant that can reach heights of ten feet. This yam, a member of a family of climbing and trailing plants that produce a variety of tubers, can reach a length well over three feet. Very narrow at one end, they ultimately swell on the far end to give a club-like appearance. Normally the plant produces two or three, ivory-colored, brittle tubers that can be left in the ground and allowed to increase in size over two or three seasons. After harvesting, the tubers can be peeled and cooked in the same way as potatoes: baked, boiled, roasted, or sliced and fried. Not grown commercially for the U.S. market.

Yams Louisiana: Although called a yam in the United States, it is really a sweet potato and is identified by its coppery skin, reddish-orange flesh, and its moist, mellow-sweet flavor. For more information see sweet potato or boniato.

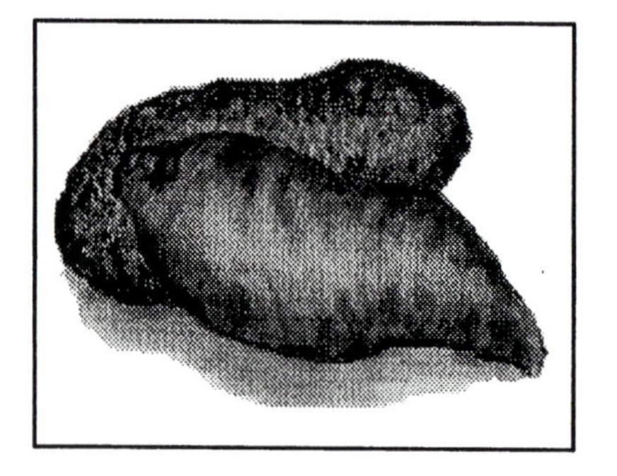

Yams: It is also known as name or true yam. What has been called a yam in the United States is actually a sweet potato, a misnomer that has existed since slaves arrived and used a word like yam to designate the American vegetable. No one outside the United States calls sweet potatoes "yams." Although there are about 600 species of yams, the one most likely to be encountered in the United States is a brown, black-brown, or orange-tan shaggy-coated tuber that may be shaped like a log. In the Pacific Island of Ponape, the size of yams are designated two-man, four-man, or six-man, describing the number needed to lift the tuber. Not an exaggeration, since 600 pound yams, six foot in length have been recorded. The raw flesh is crisp, mucilaginous, and either white, ivory, or yellow. A cooked yam's flavor is potato-like, the texture is looser, coarser, and drier, and the flavor more bland. It is boiled, baked, fried as chips or fritters, and served scalloped, creamed, or souffléd. Yams are an excellent source of potassium and zinc and contain about 80 calories per cup. Available year-round.

Many tuberous members of the species Manihot esulenta, which includes manioc, yuca, aypu, cassava, and boniata, are edible, but some, like bitter cassava, require long cooking or special preparation in order to kill toxins in them. If unknown, one should always inquire about preparation prior to using.

Appendix

FRUIT YIELDS

An important area of consumer information is that of fruit yields. General guidelines appear below for a few items subject to consumer inquiries.

Apples:	2 1/2 to 3 lbs per quart of applesauce
Avocado:	1/2 avocado per serving
Berries:	1 qt equals 3 3/4 cups and 4 to 6 servings
Cantaloupe:	1 melon yields 2 servings
Cherry Sweet:	1 lb equals 2 3/4 cups and serves 4 to 5
Cranberry:	1 lb equals 1 quart for 3 to 3 1/2 cups sauce
Grapes:	1 lb yields 3 to 4 servings
Honeydew:	1 melon yields 6 to 8 servings
Lemon:	1 lemon yields 3 to 4 tablespoon juice
Lime:	1 lime yields 1 to 2 tablespoon juice
Nectarines:	1 lb, sliced, equals 2 to 3 servings
Oranges:	1 lb yields 1 cup of juice
Papaya:	1 papaya yields 2 servings
Peach:	1 lb, sliced, equals 2 to 3 servings
Pears:	1 lb, sliced, equals 2 to 3 servings
Pineapple:	1 fruit yields 6 to 8 servings
Watermelon:	1 lb per serving

Source: Produce Marketing Association

VEGETABLE YIELDS

Listed below are some of the vegetable items subject to consumer inquiries about yields.

Asparagus:	1 lb yields 2 servings
Bean Green:	1 lb yields 3 1/2 cups and 4 to 5 servings
Bean Lima:	1 lb yields 1 cup or 2 servings.
Beets:	1 lb, diced or sliced, equals 1-3/4 cups and 3 to 4 servings
Broccoli:	1 lb serves 3 to 5, according to serving size
Brussels Sprt:	1 lb serves 4
Cabbage:	1 lb, shredded, yields 4 cups and serves 8 1 lb cooked makes 2 to 3 cups, serves 4 to 5
Carrots:	1 lb yields 4 servings
Cauliflower:	1 lb serves 4 to 5
Celery:	1 lb yields 4 serving hearts and 1 cup diced
Corn:	1 dozen ears serves 4 to 6, with 3 cups of kernels if cut from cob
Eggplant:	1 lb yields 3 cups diced and 3 to 4 servings
Greens:	1 lb beet greens, swiss chard, collards, kale, mustard greens, spinach, or turnip greens yields 2 cups when cooked and serves 3 to 4
Kohlrabi:	1 lb yields 2 cups diced, and serves 3 to 4
Mushrooms:	1 lb serves 5 to 10 when cooked as meat garnish or 8 to 14 when sliced in salads
Nappa:	1 lb yields 4 to 5 servings
Okra:	1 lb, sliced, makes 2-3/4 cups and serves 4 to 6
Onions/Leeks:	1 lb serves 4 to 5 as a cooked vegetable
Parsnips:	1 lb serves 4 sliced or cut into strips and cooked
Peas:	1 lb yields 1 1/8 cups and serves 2
Potatoes Swt:	1 lb serves 4
Potatoes:	1 lb, peeled and cubed, makes 2 1/3 cups and serves 2 to 3 2 lbs, cooked and mashed, serves 4 to 5
Rhubarb:	1 lb makes 4 to 6 servings
Salsify:	1 lb yields 2 cups diced and serves 4 to 5
Squash Hard:	1 lb yields 1 cup and serves 2
Squash, Soft:	1 lb yields 1 1/2 cups and serves 2
Turnip/Bago:	1 lb serves 2 (creamed), 4 (mashed with potatoes)

Source: Produce Marketing Association

ENVIRONMENTAL CONDITIONS

Produce	Post-Harvest	Best	
Item	**Life (Days)**	**Temp-erature**	**Humid-ity**
Apples	90-240	30-40 F	90-95%
Apricots	7-14	32 F	90-95%
Asian Pears	150-180	34 F	90-95%
Atemoyas	28-42	55 F	85-90%
Avocado, Cold Tolerant	14-28	40 F	85-90%
Cold Intolerant	14-28	55 F	85-90%
Babaco	7-21	45 F	85-90%
Bananas	7-28	56-58 F	90-95%
Barbados Cherries	49-56	32 F	85-90%
Blackberries	2-3	31-32 F	90-95%
Black Sapotes	14-21	55-60 F	85-90%
Blood Oranges	21-56	40-44 F	90-95%
Blueberries	10-18	31-32 F	90-95%
Breadfruit	14-56	55-50 F	85-90%
Cactus Pears	21	36-40 F	90-95%
Caimitos	21	38 F	90%
Calamondin	14	48-50 F	90%
Canistel	21	55-60 F	85-90%
Cantaloupes	10-14	36-41 F	95%
Carambola	21-28	48-50 F	85-90%
Cherimoyas	14-28	55 F	90-95%
Cherries, Sweet	14-21	30-31 F	90-95%
Clementines	14-28	40 F	90-95%
Coconuts	30-60	32-35 F	80-95%
Cranberries	60-120	36-40 F	90-95%
Custard Apples	28-42	41-45 F	85-90%
Durian	42-56	30-42 F	85-90%
Feijoas	14-21	41-50 F	90%
Figs, Fresh	7-10	31-32 F	85-90%
Granadillas	21-28	50 F	85-90%

Produce	Post-Harvest	Best	
Item	Life (Days)	Temperature	Humidity
Grapefruit, Az/Ca	28-42	58-60 F	85-90%
Grapefruit, Fl/Tx	28-42	50-60 F	85-90%
Grapes	56-180	32 F	85%
Guavas	14-21	41-50 F	90%
Honeydew, C_2H_4 Treated	21-28	45-50 F	90-95%
Jaboticaba	2-3	55-60 F	90-95%
Jackfruit	14-42	55 F	85-90%
Jaffa Oranges	56-84	46-50 F	85-90%
Kiwano	180	50-60 F	90%
Kiwifruit	28-84	32 F	90-95%
Langsat	14	52-58 F	85-90%
Lemons	30-180	50-55 F	85-90%
Limes	21-35	48-50 F	85-90%
Longans	21-35	35 F	90-95%
Loquats	21	32 F	90%
Lychees	21-35	35 F	90-95%
Mamey Sapotes	14-42	55-60 F	90-95%
Mangoes	14-25	55 F	85-90%
Mangosteen	14-28	55 F	85-90%
Melons, Mixed	14-21	45-50 F	90-95%
Nectarines	14-28	31-32 F	90-95%
Oranges, Az/Ca	21-56	32-48 F	85-90%
Oranges, Fl/Tx	56-84	32-48 F	85-90%
Papayas	7-21	45-55 F	85-90%
Passion Fruit	21-35	45-50 F	85-90%
Peaches	14-28	31-32 F	90-95%
Pears	60-90	32 F	90-95%
Pepino Melon	30	40 F	85-90%
Persimmons	90-120	30 F	90%
Pineapple, Mature Green	14-36	50-55 F	85-90%

Produce	Post-Harvest	Best	
Item	Life (Days)	Temp-erature	Humid-ity
Pineapple, Ripe	14-36	45 F	85-90%
Plantains	7-35	55-58 F	90-95%
Plums/Prunes	14-28	32 F	90-95%
Pomegranates	60-90	41 F	90-95%
Pummelos	84	45-48 F	85-90%
Quinces	60-90	31-32 F	90%
Raspberries, Black	2-3	32 F	90-95%
Raspberries, Red	2-3	32 F	90-95%
Sapodillos	14-21	60-68 F	85-90%
Soursop	7-14	55 F	85-90%
Strawberries	5-10	32 F	90-95%
Sugar Apples	28	45 F	85-90%
Tamarillos	70	37-40 F	85-95%
Tamarindos	21-28	45 F	90-95%
Tangerines	14-28	40 F	90-95%
Ugli Fruit	14-21	40 F	90-95%
Watermelon	14-21	50-60 F	90%
White Sapotes	14-21	67-70 F	85-90%

Source: Produce Marketing Association

NUTRITION–FRUIT

Percent of Minimum Daily Values

Item	Servings	Calories	Fat	Sodium	Vitamin A	Vitamin C	Calcium	Iron	Protein	Carbo-hydrates
Apples	1 med.	80	0.5g	0mg	0%	6%	0%	0%	0g	22g
Apricots	3 med.	60	1g	0mg	45%	20%	2%	2%	0g	11g
Avocado	2 tbs., mashed	60	6g	0mg	0%	4%	0%	0%	1g	2g
Banana	1 med.	120	1g	0mg	0%	15%	0%	2%	1g	32g
Blackberries	1 cup	60	1g	0mg	0%	50%	4%	4%	1g	12g
Blueberries	1 cup	100	1g	0mg	0%	15%	0%	2%	1g	27g
Cantaloupe	1/4 med.	50	0g	35mg	80%	80%	2%	2%	1g	13g
Carambola	1 each	40	1g	0mg	0%	30%	0%	0%	0g	8g
Cherries	1 cup	90	1g	0mg	0%	8%	0%	0%	1g	23g
Dates	5-6 dates	120	0g	0mg	0%	0%	2%	2%	1g	31g
Grapefruit	1/2 med.	70	0g	0mg	10%	80%	4%	0%	1g	18g
Grapes	1 1/2 cup	85	0g	0mg	2%	8%	2%	2%	1g	24g
Honeydew	1/10 med.	50	0g	45mg	0%	40%	0%	2%	1g	14g
Kiwifruit	2 med.	100	1.5g	0mg	2%	230%	6%	4%	5g	25g
Lemon	1 med.	20	0g	10mg	0%	35%	2%	0%	0g	6g
Lime	1 med.	20	0g	0mg	0%	35%	2%	2%	0g	7g
Mango	1/2 med.	70	0.5g	0mg	40%	15%	0%	0%	0g	17g
Nectarine	1 med.	70	1g	0mg	20%	10%	0%	0%	1g	16g
Orange	1 med.	80	0g	0mg	0%	120%	4%	0%	1g	21g
Papaya	1/2 med.	70	0g	10mg	8%	150%	4%	2%	0g	19g
Peach	2 med.	70	0g	0mg	20%	20%	0%	0%	1g	19g
Pear	1 med.	100	1g	0mg	0%	10%	2%	2%	1g	25g
Pineapple	2 slices 3”	70	0g	10mg	0%	25%	0%	0%	0g	17g
Plums	2 med.	70	1g	0mg	9%	20%	0%	0%	1g	17g
Raspberries	1 cup	50	0g	0mg	0%	40%	2%	2%	1g	17g
Strawberries	8 med.	70	0.5g	0mg	0%	130%	2%	2%	1g	17g
Tangerine	1 med.	45	1g	5mg	0%	40%	4%	0%	0g	16g
Watermelon	2 cups, diced	90	0g	10mg	10%	25%	0%	2%	1g	23g

Percent of Minimum Daily Values are based on a 2,000-calorie diet.
Source: Produce Marketing Association *“Labeling Facts”*

NUTRITION–VEGETABLES

Percent of Minimum Daily Values

Item	Servings	Calories	Fat	Sodium	Vitamin A	Vitamin C	Calcium	Iron	Protein	Carbo-hydrates
Artichokes	1 each	25	0g	70mg	2%	10%	2%	2%	2g	6g
Asparagus	5 spears	20	0g	0mg	10%	10%	2%	2%	2g	5g
Beans, Green	3/4 cup	25	0g	0mg	2%	8%	4%	2%	1g	5g
Beets	1 med.	50	0.5g	150mg	0%	4%	0%	0%	1g	11g
Broccoli	1 med. stalk	50	0.5g	70mg	10%	200%	6%	4%	4g	9g
Brussels Sprout	4 sprouts	40	0.5g	25mg	8%	120%	2%	0%	2g	6g
Cabbage	1/2 med.	25	0g	25mg	0%	40%	4%	0%	1g	6g
Carrots	1 med.	40	0g	50mg	330%	8%	2%	0%	1g	9g
Cauliflower	1/6 head	25	0g	40mg	0%	100%	2%	2%	2g	5g
Celery	2 stalks	25	0g	125mg	2%	10%	4%	2%	1g	5g
Corn, Sweet	1 med. ear	75	1g	15mg	5%	10%	0%	3%	3g	17g
Cucumbers	1/3 med.	15	0g	0mg	4%	8%	2%	2%	1g	3g
Eggplant	1/5 avg.	25	0g	0mg	0%	2%	0%	2%	1g	5g
Endive	3/4 cup, chopped	10	0g	30mg	0%	0%	2%	2%	1g	2g
Garlic	1 clove	5	0g	0mg	0%	2%	0%	0%	0g	1g
Lettuce, Head	1/6 med.	20	0g	10mg	2%	4%	2%	0%	1g	3g
Lettuce, Leaf	1 1/2 cup, shredded	15	0g	40mg	30%	4%	2%	0%	1g	3g
Mushrooms	5 med.	20	0g	0mg	0%	2%	0%	2%	2g	3g
Okra	6 pods	30	0g	15mg	10%	20%	6%	4%	2g	6g
Onions, Bulb	1 med.	60	0g	5mg	0%	15%	4%	2%	1g	16g
Onion	1/4 cup, chopped	5	0g	0mg	2%	20%	0%	4%	0g	1g
Peppers, Bell	1 med.	30	0g	0mg	6%	150%	0%	2%	1g	7g
Pepper, Chile	1 each	20	0g	10mg	80%	170%	0%	0%	1g	3g
Potatoes	1 med.	120	0g	5mg	0%	40%	0%	2%	3g	27g
Radishes	7 med.	20	0g	35mg	0%	30%	0%	0%	0g	3g
Rutabagas	1/2 cup	25	0g	15mg	8%	30%	0%	0%	1g	5g
Spinach	1 1/2 cups, shredded	40	0g	160mg	70%	25%	6%	20%	2g	10g
Squash, Summer	1/2 med.	20	0g	0mg	4%	25%	2%	2%	1g	4g
Sweet Potato	1 med.	140	0g	15g	520%	50%	2%	4%	2g	32g
Tomato	1 med.	35	1g	5mg	15%	35%	0%	2%	1g	7g

Percent of Minimum Daily Values are based on a 2,000-calorie diet.
Source: Produce Marketing Association *"Labeling Facts"*

Bibliography

Many books and magazines were researched in the writing of this work. Those listed here were particularly helpful.

Awake Magazine (Watchtower and Tract Society, Inc.)

Beck. (1984). *Produce, Fruit and Vegetable Lover's Guide.* New York: Friendly Press.

Fresh Produce A to Z. (1991). Menlo Park, CA: Sunset Publishing Co.

Funk. (1968). *Words and Their Romantic Stories.* New York: Funk & Wagnalls, Publishing Co.

Graf. (1986). *Tropica.* East Rutherford, NJ: Roehrs Publishing Co.

Hargreaves. (1964). *Tropical Trees of Hawaii.* Kailua, HA: Hargreaves Publishing.

Hessayon. (1993). *The Fruit Expert.* London, England: Transworld Publishers.

Kirk. (1975). *Wild Edible Plants of Western North America.* Happy Camp, CA: Naturegraph Publishers.

Krewer, Crocker, Meyers, Bertrand, and Horton. (1993). *Minor Fruits & Nuts in Georgia.* University of Georgia, College of Agriculture and Environmental Sciences.

Larcom. (1991). *Oriental Vegetables, Complete Guide for the Gardening Cook.* Great Britain: John Murray, LTD.

Morton. (1976). *Herbs and Spices.* New York: Golden Press.

Phillips and Rix. (1994). *Vegetables.* New York: Random House Publishing.

Reich. (1992). *Uncommon Fruits Worthy of Attention.* Reading, MA: Addison-Wesley Publishing.

Schneider. (1986). *Uncommon Fruits & Vegetables.* New York: Harper & Row Publishing.

Walden. (1963). *A Dictionary of Trees.* St. Petersburg, FL: Great Outdoors Publishing.

Williams. (1986). *Florida's Fabulous Trees.* Tampa, FL: World Wide Publications.

Author Note

I would like to extend a special thank you to the following fruit and vegetable associations who supplied information for this work.

Almond Board of California	Modesto, California
California Apple Commission	Fresno, California
California Apricot Advisory Board	Walnut Creek, California
California Artichoke Advisory Board	Castroville, California
California Avocado Commission	Santa Ana, California
California Date Administrative Committee	Indio, California
California Independent Almond Growers	Ballico, California
California Kiwifruit Commission	Sacramento, California
California Pistachio Commission	Fresno, California
California Strawberry Advisory Board	Watsonville, California
California Table Grape Commission	Fresno, California
California Tomato Board	Fresno, California
Florida Department of Agriculture	Tallahassee, Florida
Florida Lime & Avocado Committee	Homestead, Florida
Florida Sweet Corn Exchange	Orlando, Florida
Florida Tomato Exchange	Orlando, Florida
Frieda's, Inc.	Los Alamitos, California
Georgia Fruit & Vegetable Commission	Atlanta, Georgia
Idaho Potato Commission	Boise, Idaho
Louisiana Sweet Potato Commission	Opelousa, Louisiana
Michigan Apple Committee	DeWitt, Michigan
Michigan Blueberry Growers Association	Grand Junction, Michigan

Monterey Mushrooms	Santa Cruz, California
Northwest Cherry Growers	Yakima, Washington
Ocean Spray Cranberries, Inc.	Middleboro, Maine
Ontario Tender Fruit Producers	Vineland Station, Ontario, Canada
Oregon-Washington Pear Bureau	Portland, Oregon
Papaya Administrative Committee	Honolulu, Hawaii
PictSweet Mushroom Farms	Salem, Oregon
Pineapple Growers Association of Hawaii	Kunia, Hawaii
Seald-Sweet Growers, Inc.	Vero Beach, Florida
Specialty Fruits (Fresh for Flavor Foundation)	Ottawa, Ontario, Canada
Sun World International Inc.	Bakersfield, California
Sunkist Growers Inc.	Van Nuys, California
Texas & Oklahoma Watermelon Assoc.	Weatherford, Texas
Washington Apple Commission	Wenatchee, Washington
Washington Potato & Onion Association	Moses Lake, Washington
Washington State Fruit Commission	Yakima, Washington
Western Mushroom Marketing Association	San Jose, California

Index